Urban Sustainability

The Urban Sustainability Book Series is a valuable resource for sustainability and urban-related education and research. It offers an inter-disciplinary platform covering all four areas of practice, policy, education, research, and their nexus. The publications in this series are related to critical areas of sustainability, urban studies, planning, and urban geography.

This book series aims to put together cutting-edge research findings linked to the overarching field of urban sustainability. The scope and nature of the topic are broad and interdisciplinary and bring together various associated disciplines from sustainable development, environmental sciences, urbanism, etc. With many advanced research findings in the field, there is a need to put together various discussions and contributions on specific sustainability fields, covering a good range of topics on sustainable development, sustainable urbanism, and urban sustainability. Despite the broad range of issues, we note the importance of practical and policy-oriented directions, extending the literature and directions and pathways towards achieving urban sustainability.

The series will appeal to urbanists, geographers, planners, engineers, architects, governmental authorities, policymakers, researchers of all levels, and to all of those interested in a wide-ranging overview of urban sustainability and its associated fields. The series includes monographs and edited volumes, covering a range of topics under the urban sustainability topic, which can also be used for teaching materials.

Ali Cheshmehzangi · Nicholas You ·
Zaheer Allam · Bao-Jie He

People-Centered Innovation

Local Actions for Sustainable Cities and Communities

Springer

Ali Cheshmehzangi
School of Architecture, Design and Planning
The University of Queensland
Brisbane, QLD, Australia

Zaheer Allam
School of Architecture, Design and Planning
The University of Queensland
Brisbane, QLD, Australia

Nicholas You
Guangzhou Institute for Urban Innovation
Guangzhou, Guangdong, China

Bao-Jie He
School of Architecture and Urban Planning
Chongqing University
Chongqing, Chong Qing, China

ISSN 2731-6483 ISSN 2731-6491 (electronic)
Urban Sustainability
ISBN 978-981-96-5229-7 ISBN 978-981-96-5230-3 (eBook)
https://doi.org/10.1007/978-981-96-5230-3

This Springer imprint is published by the registered company Springer Nature Singapore Pte Ltd.
The registered company address is: 152 Beach Road, #21-01/04 Gateway East, Singapore 189721, Singapore

We collectively dedicate this book to caring and courageous sustainability activists of the global south. There is more hope there than we can imagine.

Preface

"The greatest threat to our planet is the belief that someone else will save it".

—Robert Swan, Author

People are more important than ever in attaining sustainable development in an era of unparalleled urbanization and environmental challenges. Urban planning and environmental stewardship require creative, people-centered solutions as cities expand and their populations rise. Although policies, technologies, and economic models are frequently used to define the concept of sustainable development, in the end, people's participation, actions, and decisions determine its ultimate meaning. Whether or not we can rise to the urgent challenges of our time will depend on the creativity and collective will of individuals and communities.

A straightforward yet profound idea is at the core of sustainable development: progress must be made to meet current needs without endangering the capacity of future generations to meet their own. However, this idea cannot be implemented on its own. It necessitates a participatory strategy that centers decision-making around people, giving them the ability to actively influence how their communities and the world develop in the future. In order to achieve sustainable development, people-centered urban innovation is not only a strategy, but also a requirement.

While urban areas are the driving forces behind economic growth, they are also the centers of social inequality and environmental degradation. Numerous issues have been brought about by the quickening pace of urbanization, including pollution, resource depletion, a lack of housing supply, and inadequate infrastructure. However, these difficulties also present opportunities. Cities are in a unique position to spur the innovation required for a sustainable future because of their density of ideas and resources. However, for this to occur, urban innovation needs to be based on the requirements and goals of the local population.

Enhancing infrastructure and productivity are only two aspects of people-centered urban innovation. It is about building resilient, equitable, and livable cities. It has to do with planning urban areas with everyone's welfare—especially the most disadvantaged—in mind. This strategy acknowledges that the active involvement of the target population is essential to the success of any urban initiative. The inclusion of people is crucial for developing urban environments that are just and sustainable, whether through community-led projects, participatory planning processes, or the incorporation of local knowledge into policy-making.

The need to mitigate and adapt to climate change is one of the biggest challenges of the twenty-first century. This global crisis has social and economic undertones in addition to environmental ones. The people least responsible for climate change—developing countries, marginalized groups, and low-income communities—are the ones who suffer its effects the most. Therefore, combating climate change necessitates a people-centered strategy that acknowledges the connections between social, economic, institutional, and environmental issues.

Giving people the tools they need to take action on a local, national, and international scale will strengthen their position in the fight to save the planet Earth. It entails promoting sustainable practices, teaching people about the value of environmental stewardship, and creating a sense of ownership and responsibility for the environment. It also entails making certain that everyone has the means, chances, and resources to support sustainable development—whether via innovative approaches to environmental problems, sustainable business practices, or grassroots activism.

Furthermore, governments and businesses alone cannot spearhead the shift to a sustainable future. It necessitates a wide-ranging movement that unites people and communities in the pursuit of common objectives. Individuals are active change agents as well as passive recipients of laws and programs. We can make the world more equitable and sustainable by utilizing the power that exists within each person and within communities.

The idea of people-centered urban innovation assumes new importance in this setting. Redefining what it means to live sustainably is more important than merely making cities more livable. It is about minimizing the ecological footprint of cities while simultaneously establishing urban environments that promote the health and well-being of all citizens. It is about making sure that everyone benefits equally from urban development and that nobody is left behind.

People are at the center of this endeavor, and we must never lose sight of this as we pursue sustainable development. Their inventiveness, tenacity, and resolve will propel the advancements required to handle the difficult problems we confront. We can create a future that is just, inclusive, and hopeful in addition to being sustainable by putting people at the center of our efforts. People play an essential and indispensable role in attaining sustainable development.

Brisbane, Australia — Ali Cheshmehzangi
Tuscany, Italy — Nicholas You
Port Louis, Mauritius — Zaheer Allam
Chongqing, China — Bao-Jie He

Acknowledgements We collectively acknowledge the Guangzhou Institute for Urban Innovation and their team for their full support and dedication to making this book project happen. We thank the competition entry participants, the critique panel members, and the respected external reviewers of the Sixth Guangzhou Award for Urban Innovation held in December 2023. We acknowledge the organizing host in Guangzhou and all sponsors who dedicated time and effort to making the event a reality through the rigid process of submissions, technical committee evaluations, seminars, final assessment by jury, online voting, and the award ceremony.

We also acknowledge the subsequent work and support of team members in providing materials, collecting data, and attending events and workshops of this collaborative work. Lastly, we acknowledge contributions and full support from our team members in six different locations in Australia, China, Italy, Kenya, and Mauritius. Most important of all, we acknowledge the project owners of all 15 cities covered in this book.

Competing Interests The authors have no competing interests to declare that are relevant to the content of this manuscript.

About This Book

People-Centered Innovation: Local Actions for Sustainable Cities and Communities examines transformative urban solutions through 15 internationally recognized case studies submitted to the 6th Guangzhou International Award for Urban Innovation in 2023. It provides valuable insights into the driving forces behind sustainable urban development, aligning closely with the United Nations' Sustainable Development Goals (SDGs).

The main focus of the book is on people-centered approaches to urban innovation, demonstrating how grassroots initiatives can tackle complex urban challenges and contribute to global sustainability targets. Each case study presents innovative strategies implemented by communities worldwide, showcasing how local actions can greatly impact sustainability and resilience. The varied examples highlight a spectrum of urban issues, including affordable housing, sustainable transportation, climate adaptation, and social inclusion, reflecting the diverse nature of urban innovation.

Key Features and Benefits:

- **Varied Global Case Studies:** The book includes 15 case studies from different regions, each presenting unique solutions to urban challenges. These examples illustrate how various cultures and contexts address common problems, providing a rich array of ideas and practices. This makes it an essential resource for urban planners, policymakers, and community leaders seeking inspiration and guidance.
- **Emphasis on People-Centered Approaches:** The book emphasizes the importance of people-centered innovation, which prioritizes community involvement and local knowledge. By focusing on how local populations engage with and drive urban initiatives, the book highlights the significance of inclusivity and participatory processes in achieving sustainable outcomes.
- **Alignment with the SDGs:** Each case study is connected to specific Sustainable Development Goals, demonstrating how local actions contribute to global sustainability objectives. The book illustrates how grassroots initiatives address key areas such as poverty reduction, climate action, and sustainable cities and communities, highlighting the critical role of local efforts in achieving the broader SDG agenda.

Role of People and Local Actions in Achieving the SDGs

The significance of people and local initiatives in achieving the SDGs cannot be emphasized enough. Sustainable urban development involves meeting the needs of local populations while also addressing global goals. The examples in this book show how empowering communities to lead their own development can result in innovative solutions to urban challenges. Local initiatives can act as drivers of broader change as they can be customized to specific community needs and circumstances. These projects not only enhance residents' quality of life but also contribute to reaching global sustainability objectives. By involving communities in urban solution planning and execution, cities can form more inclusive and fair environments that support sustainable development.

Furthermore, innovation centered around people encourages a culture of cooperation and sharing knowledge, which is crucial for tackling complex global challenges. The book demonstrates how successful local actions can motivate and educate similar initiatives in other locations, creating a ripple effect that goes beyond individual cities and communities. In summary, "*People-Centered Innovation: Local Actions for Sustainable Cities and Communities*" presents a compelling examination of how local actions and community involvement drive sustainable urban development. With its diverse case studies and focus on people-centered approaches, the book offers valuable insights into how cities and communities can contribute to the SDGs and strive for a more sustainable and fair future.

15 global case study examples covered in the book are mapped below and are highlighted in Chapters 2–16.

Contents

About the Authors

Ali Cheshmehzangi is Full Professor and Head of the School of Architecture, Design and Planning (ADP) at the University of Queensland, where he leads the School and researches across the fields of architecture, design, and planning/urbanism. He has been in the World's top 2 per cent field leader since 2022, recognized by Stanford University. He is among the top 30 global scholars in the urban sustainability research area. Ali has a wealth of experience, expertise, and a strong commitment to innovation and excellence in the fields of architecture, design, and planning, the three core disciplines of his School at UQ. With a career spanning over two decades, he has made significant contributions to the academic and professional communities, with a focus on sustainable and environmentally conscious design. So far, Ali has published over 500 journal papers, articles, conference papers, book chapters, and reports. He also has 37 other academic books, three of which have received awards at the national, provincial, and municipal levels. He also has received international awards and recognition for his research on urban resilience studies and sustainability research, as well as a Vice-Chancellor's award for his impactful contribution to higher education.

Nicholas You is a veteran urban specialist and thought leader. Prior to his retirement he was the senior policy and planning advisor to UN-Habitat and the manager of the Habitat II Conference held in Istanbul in 1996. He is the founder and honorary chairman of the UN-Habitat World Urban Campaign Steering Committee, former chairman of the Assurance Group for Urban Infrastructure of the World Business Council for Sustainable Development, former Chairman of the Urban Strategy and Innovation Council for ENGIE and former co-President of the Global Cities Business Alliance.

He is currently Executive Director of the Guangzhou Institute for Urban Innovation, co-Chair of the Open Green City Lab in Shenzhen and adjunct for the Centre for Liveable Cities, Singapore. He regularly advises central and local governments, technology companies and civil society organizations on urban sustainability, urban governance and urban innovation including the nexus between water, waste, transport and energy. He also works as a strategic planning and governance advisor to many metropolitan authorities worldwide. Since the outbreak of the pandemic, he has been part of a network of professionals facilitating the exchange of positive experiences between civil society and grassroots organizations. In 2020 he co-authored several key reports that were presented to G20.

Zaheer Allam holds a Ph.D. in Humanities, a Master of Arts (Res), an MBA, and a Bachelor of Applied Science in Architectural Science from universities in Australia and the United Kingdom. He is currently Adjunct Associate Professor at the School of Architecture, Design and Planning at The University of Queensland, Australia. Based in Mauritius, he was first Chairperson of the National Youth Environment Council (NYEC) and a board member of the Mauritius Renewable Energy Agency (MARENA) and works on a number of projects on the thematic of Smart Cities and on strategies dwelling in the increasing role of technology in Culture and the Society. Zaheer is also the African Representative of the International Society of Biourbanism (ISB), member of the Advisory Circle of the International Federation of Landscape Architects (IFLA), and a member of a number of other international bodies. He is also affiliated with Urban Systems Institute, The University of Hong Kong, Hong Kong SAR, China, and Institute for Sustainable Futures, University of Technology Sydney, Sydney 2007, Australia. Zaheer holds a number of awards and commendations, including an elevation, by the President of Mauritius, to the rank of Officer of the Order of the Star and Key of the Indian Ocean, the highest distinct order of merit in Mauritius. He is the author of over 180 peer reviewed publications and author of 15 books on the subject of Smart, Sustainable and Future Cities.

Bao-Jie He is Professor of Urban Climate and Sustainable Built Environment with the School of Architecture and Urban Planning at Chongqing University, China. He is currently leading the Centre for Climate-Resilient and Low-Carbon Cities with the focus on Heat-Resilient and Low-Carbon Urban Planning and Design. He is also Adjunct Associate Professor at the School of Architecture, Design and Planning at The University of Queensland, Australia. Baojie has published more than 180 peer-reviewed papers in highranking journals and delivered more than 40 invited talks in reputable conferences/seminars. Baojie has a SCOPUS H-index of 49 (Scopus). Baojie has been involved in several large research projects on urban climate and built environment in China and Australia. Baojie has been invited to act as Associate Editor, Topic Editor-in-Chief, Leading Guest Editor, Editorial Board Member, Conference Chair, Sessional Chair, Scientific Committee by a variety of reputable international journals and conferences. Baojie received the received the Most Cited Chinese Researchers Title in 2024, Highly Cited Researcher Title (Clarivate) in 2022, 2023, and 2024, the Sustainability Young Investigator Award in 2022, the Green Talents Award (Germany) in 2021, and National Scholarship for Outstanding Study Abroad Students (China) in 2019. Baojie was ranked as one of the Top 2% Scientists by the Mendeley since 2020.

Chapter 1
People-Centered Approach, Urban Innovation, and the SDGs

Abstract In the opening chapter, the focus is on the crucial role of innovation centered around people in attaining sustainable urban development, underscoring the significance of local initiatives and community-based approaches. Through an examination of 15 international case studies submitted for the 6th Guangzhou International Award for Urban Innovation, the chapter sheds light on how cities and communities worldwide are utilizing grassroots innovations to tackle urban challenges and contribute to the Sustainable Development Goals (SDGs) set by the United Nations. Each case study presents a distinct strategy for improving urban resilience, inclusivity, and sustainability, showcasing the effectiveness of tailored solutions that cater to the specific needs and circumstances of local communities. The section emphasizes the importance of community involvement, cooperation, and flexibility as crucial elements for successful people-centered urban innovation. By drawing from these examples, the section offers practical insights for policymakers, urban planners, and community leaders, advocating for the adoption of localized approaches that prioritize the well-being and empowerment of residents. By deepening our comprehension of the impact of people-centered innovation, this section makes a valuable contribution to the global dialogue on sustainable urban development and the establishment of resilient, inclusive, and dynamic cities.

Keywords People-centered innovation · Sustainable urban development · Local actions · Community engagement · Urban resilience · Case studies · SDGs

1.1 Introduction

As cities evolve at an unprecedented pace, the integration of human-centric principles into urban planning and development has become essential for creating resilient, inclusive, and sustainable urban environments. This has been the highlight for a few years, i.e., towards people-centered approach rather than mere development. The People-Centered Approach represents a paradigm shift in how we conceive urban spaces, placing the needs, aspirations, and well-being of individuals at the forefront of

A. Cheshmehzangi et al., *People-Centered Innovation*, Urban Sustainability,
https://doi.org/10.1007/978-981-96-5230-3_1

urban design and policy (United Nations, 2021). This approach is intricately linked to the United Nations' Sustainable Development Goals (SDGs), which provide a global framework for addressing the most pressing challenges facing humanity and the planet. As we explore the intersection of these concepts, we gain insight into how urban innovation can be leveraged to foster equitable, liveable, and sustainable cities.

The United Nations' SDGs, adopted in 2015, represent a comprehensive agenda for global development, encompassing 17 goals that address a wide range of social, economic, and environmental issues. These goals serve as a blueprint for achieving a more equitable and sustainable world by 2030 (Cheshmehzangi and Dawodu, 2019). They emphasize the importance of integrating environmental stewardship, social inclusion, and economic development to ensure that progress benefits all members of society, particularly the most vulnerable. Among these goals, several are directly relevant to urban innovation and the people-centered approach, including Goal 11: *Sustainable Cities and Communities*, which aims to make cities and human settlements inclusive, safe, resilient, and sustainable.

A people-centered approach in urban innovation prioritizes the experiences and needs of individuals and communities in the design and implementation of urban projects (Cheshmehzangi and Zou, 2024). This approach seeks to enhance the quality of life for residents by ensuring that urban spaces are accessible, functional, and responsive to diverse needs (Lotfata, 2023; Moussa, 2023). It involves engaging with communities to understand their perspectives, aspirations, and challenges, and incorporating this input into the planning and development process. By focusing on the human dimension of urban environments, this approach aims to create spaces that foster social cohesion, promote health and well-being, and support economic opportunities for all.

Urban innovation, when aligned with the principles of the SDGs, has the potential to transform cities into vibrant and sustainable places that enhance the lives of their inhabitants. Innovations in technology, design, and governance can address key urban challenges such as overcrowding, environmental degradation, and social inequality. For instance, smart city technologies can improve the efficiency of urban infrastructure and services, while green building practices and sustainable transportation solutions can reduce the environmental footprint of urban areas. Additionally, inclusive urban planning practices can ensure that all residents have access to essential services and amenities, contributing to a more equitable distribution of resources and opportunities.

Embedding SDG principles into urban innovation necessitates a comprehensive and cooperative approach that involves various stakeholders, including government agencies, private sector players, NGOs, and community organizations. This multistakeholder involvement is essential for identifying local needs, devising innovative solutions, and ensuring the widespread benefits of urban development. Through building partnerships and collaborative governance, cities can leverage the collective expertise and resources needed to accomplish the SDGs and create flourishing urban environments.

Furthermore, aligning urban innovation with the SDGs can catalyze positive global transformation (Grainger-Brown et al. 2022). As cities worldwide implement people-centric strategies and sustainable practices, they contribute to the overarching objective of achieving the SDGs and tackling global issues like climate change, poverty, and inequality (Dalby et al. 2019). Urban centers, as hubs of economic activity and cultural exchange, have a significant role in shaping a more sustainable and fair future (Alcamo et al. 2020). By embracing innovative approaches that prioritize well-being and planetary health, cities can lead the way in realizing the global vision of sustainable development.

In summary, amalgamating a people-centric approach with the SDGs provides a robust framework for propelling urban innovation and fostering more inclusive, resilient, and sustainable cities. By focusing on individuals' needs and aspirations, cities can tackle critical challenges and capitalize on opportunities for positive change. The infusion of SDG principles into urban planning and development ensures that progress is made in a manner that benefits all members of society, contributing to a more equitable and sustainable future for generations to come.

1.1.1 An Overview of Our Attempt on "Harnessing Urban Innovation to Unlock the Sustainable Development Goals"

The collection "Harnessing Urban Innovation to Unlock the Sustainable Development Goals" (Cheshmehzangi et al. 2024) investigates how cities can use innovative methods and technologies to promote the United Nations' Sustainable Development Goals (SDGs). Based on successful global examples entered the 5th Guangzhou International Award for Urban Innovation, the book provided a thorough examination of how urban innovation can advance sustainability in social inclusion, environmental protection, and economic growth. Unlike this book, the previous collection was organized into multiple thematic sections, each focusing on various aspects of urban innovation and its influence on achieving the SDGs. In here, we aim for people-centered projects and examples to showcase pathways, lessons learnt, and best practices. The previous collection starts by discussing the crucial role of urban areas in the global pursuit of sustainability, emphasizing that cities are vital in addressing global challenges such as climate change, poverty, and inequality as centers of population and economic activity.

Under its four sections, case study examples of the book explored specific innovations and strategies that can assist cities in meeting the SDGs, including progress in green building technologies, smart city solutions, and sustainable transportation systems. While some projects were context-specific, they all had a range of lessons learnt that could be extracted for other cities and communities elsewhere. The projects presented examples from across the globe that demonstrated how cities have effectively employed innovative methods to enhance urban resilience, enhance quality of life, and decrease environmental impacts.

As a follow-up to our previous collection, this book places significant emphasis on integrating technological advancements with people-centered approaches, examining how digital tools, data analytics, and participatory planning can be used to create more inclusive, fair, and efficient urban environments (Dawodu et al. 2021). Regardless of the outcomes and pathways, people-centered approaches are highlighted in all cases. Moreover, the book examines the significance of multi-stakeholder collaboration in driving urban innovation and achieving the SDGs, highlighting the necessity for partnerships between government, the private sector, and civil society.

Overall, "Harnessing Urban Innovation to Unlock the Sustainable Development Goals" (Cheshmehzangi et al. 2024) offers valuable insights and practical advice for urban leaders, policymakers, and researchers. It emphasizes the potential of innovative urban solutions to contribute to a sustainable future and accomplish the global objectives outlined by the SDGs. This book can be regarded as a follow up, providing a holistic coverage of people-centered innovation in cities and communities.

1.2 Aim and Objectives of the Book

This collection explores the potential of grassroots innovations to promote sustainable urban development by addressing the specific needs and desires of local communities. Its main objective is to underscore the transformational impact of people-centered approaches in realizing sustainable urban futures (Allam and Cheshmehzangi 2024) and offers practical insights through a series of global case studies.

The primary goal of the publication is to demonstrate how locally-driven initiatives and community-based innovations can tackle critical urban challenges and contribute to the broader objectives of sustainable development. By examining a diverse array of examples, the book illustrates how customizing solutions to fit the unique context of each city or community can result in significant enhancements in sustainability, fairness, and resilience.

The objectives of this book are as follows (Fig. 1.1):

1. **Showcasing Effective Local Innovations**: It offers in-depth analyses of 15 global case studies, each showcasing successful people-centered innovations submitted to the 6th Guangzhou International Award for Urban Innovation. These cases illustrate a range of approaches to urban sustainability, from community-led green spaces to inclusive public services.
2. **Exploring the Impact of Community Engagement**: It investigates how involving local stakeholders in the planning and execution of urban projects can lead to more impactful and sustainable outcomes. The case studies provide insights into how community participation improves project relevance, effectiveness, and long-term success.

Fig. 1.1 Summary of objectives for this collection

3. **Identifying Key Success Factors**: The book identifies common factors that contribute to the success of people-centered innovations, such as collaboration, local knowledge, and adaptability. These insights aim to offer actionable lessons for other cities and communities aiming to implement similar strategies.
4. **Encouraging Global Knowledge Sharing**: By featuring a broad range of international examples, it promotes the exchange of ideas and best practices across various regions and contexts.

In summary, "People-Centered Innovation" highlights the significance of local actions and community engagement in developing, achieving, and boosting sustainable urban development, serving as a valuable resource for urban planners, policymakers, and community leaders worldwide.

1.3 Brief Summary of 15 Shortlisted Cities

In conclusion, we will provide a comprehensive list and overview of each selected project submitted to the 6th Guangzhou International Award for Urban Innovation. These projects showcase a wide variety of inventive approaches to urban challenges, demonstrating how local efforts and strategies centered around people can effectively contribute to sustainable urban development. The summaries will highlight the primary goals, results, and impacts of each project, offering valuable insights into the real-world application of community-driven innovation in different global settings. We would like to acknowledge that all materials presented have been provided by the project owners, with appropriate permissions obtained by the event organizers and the panel of experts participating in this international initiative. This compilation of projects not only bears witness to the resourcefulness and originality of urban communities worldwide but also serves as a valuable reference for policy-makers, urban planners, and researchers looking to implement similar strategies in their own cities. Through sharing these examples, our aim is to inspire further innovation and collaboration in the pursuit of people-centered innovation for achieving the Sustainable Development Goals and promoting sustainable cities and communities.

Here, is the list of all projects or initiatives, including the name of city, location, and the submitted initiatives under this competition (Table 1.1).

Table 1.1 List of all 15 shortlisted initiatives/projects submitted to the 6th Guangzhou International Award for Urban Innovation (in alphabetical order)

No	City/ local government	Initiative
1	Antalya, Türkiye	Eco-friendly farmer card: Protect your environment, Collect points, Win prizes
2	Bogotá, Colombia	Bogotá care blocks: How a city is reorganizing itself for women and achieving the SDGs
3	Cape Town, South Africa	Building resilient food systems in Cape Town
4	Gwangju, Republic of Korea	The last bastion for protecting human dignity "GwangJuness Community Care"
5	Halandri, Greece	"Cultural H.ID.RA.N.T.—cultural hidden IDentities ReAppear through Networks of Water." Ancient engineering (of sustainable water management) meets urban innovation through citizens' participation
6	Iztapalapa, Mexico	UTOPIAS—Units of transformation and organization for social inclusion and harmony
7	Jakarta, Indonesia	Reclaiming spaces for children in public transportation through @Navi_Station Initiatives
8	Kazan, Russia	Rehabilitation of the City Park "Chaikovoye Lake" in Kazan
9	Kampala, Uganda	An innovation towards pit emptying services through a GIS-enabled App (Weyonje) of Kampala Capital City Authority (KCCA)
10	Mannheim, Germany	Local SDG implementation: An annual report on the achievement of the SDGs as a management tool and the large area-transformation into a SDG-park
11	Pimpri Chinchwad, India	Navi Disha: Community toilet model
12	Ramallah, Palestine	Forum of expertise—Active Aging for a vibrant future—empowering seniors to thrive through voluntarism, lifelong learning, wellbeing, and community led initiatives
13	São Paulo, Brazil	Programa tô legal
14	Tehran, Iran	Substitution of chemical poisons by using pruning waste in the production of herbal formulations
15	Xianning, China	Enhancing traffic safety near schools through child-friendly collaborative construction and sharing approach

We also include a list of other entries to the 6th Guangzhou International Award for Urban Innovation, which are considered as the deserving initiatives (Table 1.2).

Here, we provide a summary of the 15 shortlisted initiatives that are covered in the following chapters of the book. At the end of this chapter, Boxes 1.1–1.10 provide details of the award and information associated with the shortlisted initiatives or projects.

1. Antalya, Türkiye: Eco-friendly Farmer Card—Protect your environment, collect points, win prizes

Antalya Metropolitan Municipality's program aims to minimize the harmful impact of agricultural packaging waste on the environment and public health, especially in areas with intensive farming. The project focuses on reducing plastic pollution caused by pesticide-laden packaging, which can harm soil, pollute the environment, and indirectly affect 98% of Antalya's drinking water.

Farmers can use solar-powered, fire-resistant vending machines equipped with agricultural waste collection technology by using personalized key fobs. As a reward for depositing waste, farmers receive farming sensors, greenhouse materials, and tools, while also receiving training on identifying and disposing of hazardous waste. This is especially important in a community that currently uses 10% of Türkiye's pesticides.

The waste collection and storage initiative involves collaboration between metropolitan and district municipalities, as well as cooperation between the public and citizens. In addition, there is a partnership with a local company that makes solar panel-operated waste-vending machines, equipped with impact sensors, cameras, GPS tracking, and other features. These partnerships include municipalities, citizens, the private sector, and academia, addressing agricultural packaging waste and preventing leachate through innovative coating technologies.

The initiative follows the principles of the circular economy, emphasizing sustainable resource use and waste reduction. It promotes material recycling, energy generation, reuse, alternative materials development, sustainable production, innovation, and employment opportunities. The model aims to change citizens' views from "Polluter Pays" to "Protector Wins" within the circular economy framework, challenging misconceptions about waste and responsibility.

This initiative was chosen for its significant contribution to environmental sustainability, the circular economy, local economic development, and sustainable communities, all while reducing carbon emissions. The technology has the potential to benefit countries with similar agricultural patterns. The Eco-Friendly Farmer Card initiative in Antalya exemplifies an innovative approach to addressing agricultural packaging waste and its associated environmental and health concerns, incorporating technology, incentives, and environmental stewardship to develop a more sustainable and responsible agricultural sector.

Table 1.2 List of 30 deserving initiatives/projects submitted to the 6th Guangzhou International Award for Urban Innovation (in alphabetical order)

No	City/local government	Initiative
1	Buenos Aires, Argentina	Localization and SDG voluntary local reviews of the city of Buenos Aires
2	Buenos Aires, Argentina	Ciudad 3D: a tool to promote urban development in the City of Buenos Aires
3	Berlin, Germany	Give administration a human face
4	Bontang, Indonesia	Geospatial based persons with disabilities card and services
5	Baia Mare, Romania	Baia Mare from a mining and polluted city towards a green, innovative and participatory city: SPIRE—smart post-industrial regenerative ecosystem
6	Barcelona, Spain	Climate Shelters in Schools: Adapting schools to climate change through green, blue and grey solutions
7	Daggar, Pakistan	"Building Bridges: Connecting communities for a better tomorrow"
8	Dubai, United Arab Emirates	Smart System for Infrastructure & Utilities Lands Services
9	eThekwini, South Africa	eThekwini Strat Hub
10	Guangzhou, China	Integrated and smart neonatal health management system-empowering mothers, The family and the community
11	Gimhae, Republic of Korea	Good-bye to plastic flowers and disposable dishes!—carbon neutral and Happy Green City, Gimhae
12	Gwanak-gu, Seoul, Republic of Korea	Korean Youth Capital, Gwanak-gu
13	Istanbul, Türkiye	Mucilage-Free Marmara
14	Istanbul, Türkiye	Istanbul play master plan
15	Krasnoyarsk, Russia	Svetly park as a supportive space for open communication between every social category of residents
16	Konya, Türkiye	Building Resilient Cities with a Human-Centered Approach: Konya Model
17	Medellín, Colombia	Promoting food sovereignty by strengthening food marketing through Short Marketing Circuits
18	Moscow, Russia	Leaders of digital transformation
19	Nilüfer, Türkiye	1. Nilufer food living Lab (NFLL) 2. Nilüfer social entrepreneurship center
20	Palmira, Colombia	PAZOS: Peace and opportunities for Palmira
21	Quezon City, Philippines	Quezon City's community-based urban farming program

(continued)

Table 1.2 (continued)

No	City/local government	Initiative
22	Recife, Brazil	Compaz Recife—community peace centers
23	Rotterdam, Netherlands	Growing up healthy in Rotterdam Beverwaard—a multiannual action-research to improve the lifestyle of children and their families
24	Ramallah, Palestine	1. Ramallah City stepping toward a greener Future 2. Revolutionizing Municipal Services through GIS
25	São Paulo, Brazil	URANO predictive rapid response system for urban maintenance
26	Salvador, Brazil	Saúde nos Bairros (Health in the Neighbourhoods)
27	Shanghai, China	Century-old Zhangyuan and its journey of renewal
28	Thane, India	Cluster based redevelopment of Thane through urban renewal programme
29	Tharaka Nithi, Kenya	Enhancing women's livelihoods labor force participation through the crèche program
30	Tlajomulco de Zúñiga, Mexico	Renta tu Casa (Rent Your Home)

2. Bogotá, Colombia: Bogotá Care Blocks—How a city is reorganizing itself for women and achieving the SDGs

The structural cause of gender inequality has long been attributed to women's "time poverty" due to the unequal burden of unpaid care work that falls predominantly on women, resulting in their lack of economic autonomy. Women who are primary caregivers often give up pursuing higher education and are more susceptible to chronic mental and physical health issues, as well as a higher risk of experiencing gender-based violence. In Bogotá, 30% of women are engaged in full-time unpaid caregiving, with 91% of them belonging to the low-income bracket and 33% devoid of any free time for self-care.

In 2020, Bogotá, the capital of Colombia, introduced its Care System aimed at acknowledging, reducing, and redistributing care responsibilities. The initiative primarily operates through Care Blocks that offer easy access to educational, leisure, and income-generating services for caregivers, while simultaneously providing professional assistance to the individuals being cared for, such as children, people with disabilities, and the elderly. Furthermore, a cultural change strategy called "Care School for Men" was also implemented to educate men about recognizing and sharing care responsibilities more equitably within the household. As a result, caregivers have gained an extra 4 h daily for studying, earning income, or resting, thanks to the provision of childcare and laundry services.

This initiative not only promotes gender equality, but also contributes to environmental sustainability. It serves as a model that (i) acknowledges the role of care and the invaluable contribution of caregivers to society; (ii) more equitably redistributes the responsibility of care among members of society; and (iii) reduces women's unpaid care work, allowing them to engage in education, formal employment, well-being, as well as social and political activities. Civil society participation played a crucial role in the design and implementation of Care Blocks.

Moreover, this initiative supports several Sustainable Development Goals (SDGs) including SDG 5 (Gender Equality), SDG 3 (Good Health and Well-being), SDG 11 (Sustainable Cities and Communities), SDG 16 (Peace, Justice and Strong Institutions), and SDG 17 (Partnerships for the Goals), thereby contributing to the global agenda.

3. Cape Town, South Africa: Building resilient food systems

The food systems in Cape Town are intricate and have limited governance. Recent unexpected events such as drought, COVID-19, load shedding, and a minibus taxi strike have highlighted the necessity for a food systems approach that acknowledges the growing occurrence of risks affecting food access in vulnerable communities. Food insecurity became a significant vulnerability during these crises. The initiation of the Food Systems initiative is based on the Cape Town Resilience Strategy (2019) and is part of the city's five-year development plan.

The resilient food systems initiative aims to tackle the health challenges arising from food insecurity, which lead to issues like child malnutrition or obesity, and contribute to the high prevalence of non-communicable diseases (NCDs). The majority of the work within the food systems initiative involves advocacy and knowledge building. This is accomplished through a Food Systems Working Group comprising both internal and external stakeholders that leverage knowledge, networks, and the city's efforts to achieve better food outcomes. Additionally, the city organizes a Food Indaba (meeting) internally and a Food Systems Community of Practice to coordinate and promote research.

Some of the main factors for choosing this initiative include: (i) it recognized the risks and vulnerabilities of the city immediately after the COVID-19 pandemic; (ii) it initiated the resolution of the identified key risk; (iii) the program was integrated into the city's development plan; (iv) it established partnerships to address the long-term impact of reducing non-communicable diseases occurrences.

4. Gwangju, Republic of Korea: GwangJuness Community Care

The city of Gwangju goes beyond just a slogan with its commitment to ensure no citizen is excluded. In the past, Korea's caregiving system only supported those who met certain age and income criteria, leaving many without assistance. Gwangju has introduced an innovative care model called Gwangjuness Community Care, which is open to anyone requiring care.

Care managers from the smallest administrative division, the Dong-office, visit individuals in their homes to assess their needs and apply for necessary services. Gwangju has rolled out 16 new care services and plans to invest US$7.74 million annually to encourage economically challenged citizens to utilize these services. "Care Call," a single application channel, has been introduced for added convenience. To ensure effective implementation, Gwangju has set up a dedicated division, enlisted 320 care managers, and engaged 38 private sector care service providers.

This initiative not only benefits school children but also improves pedestrian and public spaces through collaborative planning and design, leading to enhanced community well-being and increased local retail activities across all age and income groups. The initiative directly supports SDG 11 (Sustainable Cities and Communities) and SDG 3 (Good Health and Well-being).

5. Halandri, Greece: Cultural H.ID.RA.N.T. (Cultural Hidden IDentities ReAppear through Networks of Water)—Ancient engineering of sustainable water management meets urban innovation through citizens' participation

The goal of this project in Greece is to activate the local hidden heritage and resources to improve the well-being of citizens. The focus is on reintroducing the Hadrians aqueduct, an underground Roman-era structure, to city life in a way that goes beyond traditional "sightseeing" to showcase and communicate its cultural heritage, ancient engineering techniques, and its role as a historic monument, modern landmark, unused water source, and urban infrastructure component.

In addition to repositioning the aqueduct, the project promotes environmentally friendly urban policies and sustainable water management. It aims to create new high-quality green public spaces by revitalizing four neighborhoods in Halandri. Furthermore, the still functional aqueduct will be repurposed for irrigation, reducing the previous reliance on potable water. Through inclusive processes, Halandri communities are revitalized by engaging with their heritage and shared water, leading to a stronger local economy based on internal, resilient development rather than relying on tourism, which strains the city's resources and undermines community solidarity.

The initiative's uniqueness lies in its multifaceted solutions to various issues. It has enhanced the quality of public green spaces, emphasized the city's cultural heritage and community spirit, built a citizen-focused sustainable water infrastructure following a circular economic model, and established participatory processes and institutions to manage urban resources.

It has been shortlisted due to its innovative approach to improve the living environment for residential communities and promote local agriculture and tourism by utilizing currently neglected assets. It integrates cultural, social, environmental, and economic sectors through a cross-sectoral approach. The initiative demonstrates a circular economy model by involving participatory planning and governance. Above all, it empowers communities to take control of public assets to bring about creative heritage regeneration.

The project is set to be implemented over a four-year period. The participatory urban planning HIDRANT festival and digital local history platform have already been completed and are in operation. The Halandri Adrian Community Network is scheduled to be established by December 2023, while the regeneration works and non-potable water network will be completed by April 2024, when water delivery will begin.

6. Iztapalapa, Mexico City, Mexico: UTOPIAS—Urban transformation and organization for social inclusion and harmony

Iztapalapa, located in Mexico City, is home to nearly 2 million residents, making it the most densely populated municipality in the region. A significant 43% of the population is living in poverty, with 33% of residents falling between the ages of 15 and 34, and 23% working in the informal economy. One of the primary challenges faced by the local population is the presence of socio-economic and territorial inequalities.

The UTOPIAS initiative has been implemented as a means to address these challenges by focusing on social and urban transformation, aiming to reduce socio-urban inequalities and violence. This initiative is founded on principles of the "right to the city," seeking to achieve this through the revitalization of deteriorated public spaces and the development of a comprehensive system of large-scale public facilities. These facilities are designed to promote the health and well-being of underprivileged communities. The primary aim is to ensure equitable access to public spaces and facilities, offering free social, cultural, sporting, and recreational activities, as well as social services to individuals of all age groups and genders. Moreover, particular attention is devoted to providing activities for vulnerable groups such as the elderly, individuals with disabilities, women who are victims of abuse and violence, people struggling with addiction, and at-risk youth.

The initiative is characterized by an innovative approach to urban management, guided by an integrated, participatory, trans-disciplinary, and trans-sectorial vision. As a result, it has already revitalized and transformed over 540,000 square meters of degraded public spaces, establishing 12 UTOPIAS in just 4 years. These developments have the capacity to positively impact the lives of over 100,000 people who utilize the facilities on a weekly basis. Furthermore, the initiative underlines Iztapalapa's dedication to advancing Sustainable Development Goal 3 (Good Health and Well-being), SDG 5 (Gender Equality), SDG 10 (Reduced Inequalities), SDG 11 (Sustainable Cities and Communities), as well as SDG 16 (Peace, Justice, and Strong Institutions).

Jakarta, Indonesia: Reclaiming spaces for children using public transportation through the Navi Station initiative

In Jakarta, there is an effort to prepare the younger generation for safe use of rail-based public transportation, which involves developing a toolkit tailored for children aged 5–7 to enhance their understanding of train stations. The toolkit includes toys like Lego, activity books, playmats, and papercraft featuring rail transport logos and mascots. Its content and approach have been endorsed by pre-school teachers and are evaluated through pre-tests, post-tests, and edu-visits (observation). This initiative involves collaboration among the Bandung Institute of Technology, the Jakarta Transportation Authority, and rail operators, with the Ministry of Education also supporting its implementation. It serves as a great model for promoting universal accessibility to public facilities, including transportation, by empowering children to navigate train stations.

7. Kazan, Russian Federation: Park to rest and recuperate

Amid the COVID-19 pandemic, the hospital near Chaikovoye Lake in Kazan realized the absence of outdoor spaces for patients to recover and for medical staff to unwind. The neglected hospital surroundings and heavily polluted lake prompted the development of a versatile public park. The aim was to create a multifunctional park for rehabilitation, recreation, and preservation of the lake's ecosystem. The project involved close collaboration with relevant ministries, health industry representatives, hospital employees, patients, and local residents, who actively participated in planning, designing, and implementing the initiative. The park was partially landscaped in 2021 and fully operational the following year. It is equipped to accommodate people with disabilities or limited mobility and is managed by the Directorate of Parks and Squares in Kazan. The park and the now pristine lake have yielded significant social, health, and environmental benefits. Patients can engage in therapies and enjoy the scenic surroundings, while carefully selected plants serve therapeutic purposes. Hospital staff now have a space for relaxation and sports activities. The rehabilitated lake and selected aquatic plants contribute to its self-cleaning. The park has also become a center for social interaction, hosting public events and classes. Local residents now have an eco-friendly environment near the lake. As a result of the initiative, economic activities in the surrounding areas have increased. This initiative offers a sustainable environmental solution to a health concern and contributes to several Sustainable Development Goals, including Good Health and Well-being, Clean Water and Sanitation, Decent Work and Economic Growth, Industry, Innovation and Infrastructure, and Sustainable Cities and Communities.

8. Kampala, Uganda: An innovation for pit-emptying services using a GIS-enabled application

Kampala, the capital city of Uganda, has a population of 1,738,000 and experiences an urbanization rate of 5.6% annually. More than 60% of the population resides in informal housing, and only 10–15% of the city has access to the formal sewerage system. As a result, 90% of the population depends on on-site sanitation, leading to a necessity for waste-emptying services for many residents. Unfortunately, vendors often dispose of waste from pit latrines and septic tanks haphazardly into the environment.

To address these challenges, the Weyonje App was developed and launched. This mobile app enables citizens to request emptying services, which are monitored until the sludge is safely transported to the regulated dumpsite. Those without smartphones, particularly residents of informal settlements, can seek help from the Village Health Team (VHT), which uses the Weyonje Village Health Team Application to place service requests on their behalf. The app ensures the safe transport and management of faecal sludge at the treatment plant, with data from the app being incorporated into the city's data management dashboard to monitor progress and identify service gaps.

This initiative was chosen for its potential to bring about positive behavioral change, enhance environmental practices, and prevent public health crises. Notably, the initiative was lauded for its inclusivity of the urban poor, especially those without access to mobile phones. It represents an innovative data management solution that creates local employment opportunities through the Weyonje Village Health Team.

9. Mannheim, Germany: Annual report on the achievement of the SDGs as a management tool and transformation of vacant land into a park

The City of Mannheim in Germany combined two interconnected projects into a single submission. Since 2015, the city has been actively tracking progress towards the SDGs, but in 2022, it enhanced its monitoring process by deciding to present the city's annual Voluntary Local Review report as a key municipal instrument, alongside the mission statement, budget plan, and major projects. (VLR reports are reports by sub-national governments on their progress on the SDGs.) By incorporating the VLR into the city's regular management system through an annual report, Mannheim has embedded the SDGs into all aspects of the city's management and expanded on previous reviews.

Integrating the VLR into the official management tool enables the administration, councilors, and the local community to better understand how the city's actions align with the SDGs. The decision to publish a document that translates global commitments into local actions through a binding resolution promotes transparency and accountability. Furthermore, it allows councilors to focus their budget decisions on identifying gaps and needs. This action also aligns with Mannheim's commitment to global responsibility for advancing city-to-city learning and encouraging other cities to localize global goals.

The second project entailed organizing a major event centered around the SDGs as part of the redevelopment of a large urban space left vacant after the departure of a US military base. Citizens were consulted on how to repurpose the former military base, leading to the decision to transform it into a community park. Social inclusion and gender were prioritized in the park's design.

The decision to host a significant Gardening Show known as BUGA (facilitated by the Federal Government) specifically focused on the SDGs was initially made through a citizens' referendum and later confirmed by resolution by the Municipal Council. The strategy was then developed and presented to the council in 2017. The Gardening Show took place in 2023.

Plans are now underway to convert the 80-hectare site into a landscaped park and recreation area that will serve as a habitat for diverse species and a venue for sports and other recreational activities. The BUGA Park is expected to decrease the city's temperature by an average of 1.5 degrees and will continue to provide a source of fresh air. It is anticipated that the public park will improve the quality of life for Mannheim's residents and those in the metropolitan region, which is home to 2.3 million people. Approximately 2 million visitors are projected to visit the park for the BUGA every two years.

Mannheim's BUGA is dedicated to promoting environmental and climate protection, conserving resources, and ensuring food security. The project also showcases exhibits on local urban food production and clean, climate-friendly energy generation in cities. The initiative involves a nonprofit limited liability company collaborating with politicians, urban planners, landscape architects, citizens, and other stakeholders. The exhibition area will host over 5000 events related to themes like "New Mobility for Families," "Climate-Neutral Mobility—Flying with Fuel Cells," and "Insect-friendly Green Care," providing a platform for organizations, government entities, and initiatives focused on the SDGs. BUGA 23 Mannheim holds the distinction of being the first city in Germany to achieve CO^2-neutral certification and address climate, environmental, and species protection issues directly. The initiatives undertaken by the city of Mannheim serve as clear examples of how a city can integrate the SDGs into its governance processes, including budgeting and annual reporting. These initiatives were chosen for their demonstration of the city's leadership in improving the urban environment, providing valuable insights for other local authorities on incorporating VLR into decision-making. The city has effectively integrated the SDGs into its governance processes, thereby increasing transparency and community awareness. The physical project implementation reflects the commitment of the community and council towards achieving the SDGs and offers practical lessons for municipalities worldwide, inspiring similar efforts.

10. Pimpri Chinchwad, India: Navi Disha community toilet model

In the Indian city of Pimpri Chinchwad, access to sanitation facilities for families living in informal settlements is limited. To address this issue, the municipality has taken the initiative to build 160 community toilets in 71 slums. The operation and maintenance of these community facilities are efficiently managed by entrusting the responsibility to the communities themselves. As part of the Navi Disha program, women from these communities have formed groups to oversee the operation and maintenance of the community toilets. The idea for this initiative was developed through consultations with the communities, where the members identified the challenges in operation and maintenance and suggested assigning this responsibility to women's groups. These women's groups have been provided training in community toilet management using a capacity-building approach through trainer training. To ensure the smooth operation and upkeep of the facilities, the sanitation inspector from the municipality, along with 17 social workers, has been trained, and they have subsequently trained members of the women's groups. Additionally, the municipality provides a monthly payment to each women's group to cover the costs of sanitation equipment and cleaning materials. Implementing the user-pays principle, the initiative aims to generate revenue streams for the women's groups. This initiative serves as a notable example of integrating gender considerations throughout the sanitation value chain and of fostering an effective partnership between the public and the community. Apart from reimbursing costs and providing training, the municipality also offers management support to the women's groups, providing them with comprehensive operating manuals, simplified procedures, and efficient monitoring systems.

11. Ramallah, Palestine: Forum of Expertise, a community-led active aging initiative that empowers seniors to thrive through voluntarism and life-long learning

"Forum of Expertise" is a cutting-edge community hub located in Ramallah, with a focus on enhancing the lives of older individuals through the provision of a space for social interaction, participation, and overall well-being. The center offers four main programs: an intergenerational platform for the exchange of knowledge and expertise, a customized process for voluntary involvement, a targeted capacity-building program, and comprehensive well-being programs that cater to various aspects of older individuals' lives.

Innovation is evident across physical, programmatic, and operational aspects. The Forum establishes a physical environment that is inclusive to all, facilitating formal and informal gatherings to promote interaction and address social isolation among older individuals. A variety of programs are developed with input from the community. The Forum has adopted a decentralized structure that involves the community in decision-making, oversight, and assessment, with regular feedback loops to support ongoing and continuously evolving processes and strategies.

The initiative addresses a widespread issue and is dedicated to utilizing the skills and experiences of older members of society to participate in a diverse range of activities aimed at combating social loneliness and encouraging intergenerational engagement in innovative ways. The initiative is designed to ensure the ongoing adaptability of its programs in response to changing circumstances, embracing trial-and-error and incorporating robust evaluation and feedback from the community. Throughout the COVID-19 pandemic, the Forum responded to the challenges of isolation with a series of innovative online initiatives. These initiatives included a documentary program comprising of 35 episodes, featuring Palestinian musicians exploring diverse musical styles and traditions. Additionally, the Forum conducted 18 in-depth interviews with experienced creative writers, offering insights into their journeys. These programs were made accessible online and via YouTube, reaching a broad audience and promoting engagement among older individuals during the lockdown. Its impact indicators also demonstrate innovative and holistic short- and long-term approaches. Its potential to be implemented in other neighborhoods and cities is evident. It has the potential to become a transformative model with relevance beyond the immediate community.

This initiative particularly aligns with SDG 3 (Health and Well-being), SDG 4 (Quality Education), and SDG 11 (Sustainable Cities and Communities).

12. São Paulo,, Brazil: Programa Tô Legal

Tô Legal aims to simplify and speed up the process of granting permits and authorization to individuals working informally in public areas and sidewalks. It provides an opportunity for those unable to secure formal employment to work legally, fostering entrepreneurship and improving transparency, efficiency, and location monitoring through a electronic system funded by and used in São Paulo, Brazil. This electronic system allows citizens working illegally on the streets to transition to formal work gradually and could potentially lead to establishing registered companies or becoming microentrepreneurs. Tô Legal expedites the issuance of Authorization Ordinance for Commerce and Service Provision (PCS) within four days after application via a graphical representation of locations and automated approvals. This authorization allows individuals to engage in economic activities in public areas or sidewalks, promoting entrepreneurship among unemployed, youth, and elderly workers lacking job opportunities. Additionally, the Tô Legal portal offers users information on other vendors selling products on a given street, enabling users to make more informed choices about their business location and providing access to other municipal services in one place. Street entrepreneurs' lives have been improved as they no longer need to visit a sub-prefecture office to apply for documents, update their registration, or pay municipal fees since these services are now available online. Civil servants can easily access permit and authorization information, expediting inspection procedures, and streamlining revenue collection channels, directing funds directly to the Municipal Treasury. This initiative represents an innovative governance approach utilizing technology to provide pathways for potential formalization and strengthen the agency of those working in the informal sector. Tô Legal contributes to the global urban agenda by providing legal identity and standing to urban residents, strengthening economic and social rights in line with the New Urban Agenda and SDGs 8, 9, 10, and 11.

13. Tehran, Iran: Substituting chemical poisons with pruning waste in the production of herbal formulations

The introduction of pruning plants and Egyptian silk plants marks a significant shift in replacing chemical pesticides in the production process of herbal formulations. This Tehran-based initiative is driving the transformation of modern cities from their industrial past to more environmentally friendly urban spaces. The project utilizes urban green areas to combat urban pests and eliminates the reliance on harmful chemical pesticides, leading to cost savings. The key focus of this undertaking is the use of natural resources, particularly pruning plants and Egyptian silk plants, in creating herbal formulations. Extensive research has been conducted to confirm their effectiveness, resulting in the publication of five scientific papers. This project is at the forefront of promoting a natural, eco-friendly approach to herbal formulation production, contributing to the transition of modern cities into toxin-free, nature-friendly urban environments. The project is aligned with ten Sustainable Development Goals (SDGs), making it an essential contributor to the global movement for a more sustainable and environmentally responsible future.

14. Xianning, Hubei Province, China: Enhancing traffic safety near schools through a child-friendly collaborative construction and sharing approach

In Xianning, China, the lack of proper municipal infrastructure, such as sidewalks and traffic signals, resulted in frequent conflicts between vehicles and pedestrians. One common issue was the "school run," where parents or guardians drove children to school due to safety concerns for pedestrians. To address these challenges and ensure the safety of children traveling to and from school, Xianning introduced a comprehensive "government + enterprise + community" collaboration program focused on enhancing infrastructure through participatory planning and design. This effort, led by the Xianning municipal government, involved seven municipal departments, design institutes, school teachers, parents, the local community, local estate management companies, and the students themselves. Using approaches like the "franchise model," land use adjustments, segregating pedestrian and vehicular traffic, implementing walking-bus routes, and creating designated waiting areas, the initiative successfully transformed the school surroundings into safe and high-quality communal spaces and developed shared gardens. Additionally, a series of environmental improvement activities for students were organized to enhance the school's surroundings, creating ideal campus conditions and safe pathways for children.

Since the initiative's launch, there has been a significant impact: the percentage of students walking to school has surged from 30 to an impressive 95%. Volunteers oversee designated walking routes for parents, and effective traffic management has reduced the annual number of traffic accidents from over 50 to just one in 2023. The roads leading to the school have been completely revamped to separate vehicular and pedestrian traffic, stimulating the opening of retail stores and dining establishments that benefit the local community. An estate management company also contributed by providing temporary parking facilities and dismantling a perimeter wall to open up the "last mile" of the pedestrian walkway and provide students with a view of the estate's private gardens. The success of the initiative has prompted Xianning authorities to expand the approach to 20 other schools and has inspired similar initiatives in Hubei Province and the City of Guangzhou. It has also earned international recognition from UN-Habitat and UNICEF as a best practice in promoting child-friendly cities.

The youth-centered approach to urban safety and place-making demonstrated by this initiative is an excellent model for combining efforts to enhance traffic safety, revitalize urban areas, create high-quality public spaces, and encourage people to see the city from a child's perspective. This initiative effectively integrates these objectives and seizes the opportunities presented by traffic safety improvement projects to create successful public spaces for children. It promotes their education and interaction with parents and neighbors while fostering a sense of community ownership and pride. With low investments and a short implementation time, this initiative is inherently replicable and scalable. Its comprehensive approach, which goes beyond mere infrastructure, and the active involvement of local stakeholders, are also noteworthy. Furthermore, the project aligns with SDG 4 (Quality Education) and SDG 11 (Sustainable Cities and Communities).

Box 1.1—Brief Introduction to the Guangzhou Institute for Urban Innovation

The Guangzhou International Award for Urban Innovation (the Guangzhou Award) is co-hosted by the United Cities and Local Governments (UCLG), the World Association of the Major Metropolises (Metropolis) and the City of Guangzhou. The aim of the Guangzhou Award is to recognize innovation in improving social, economic and environmental sustainability in cities and regions and, in so doing, to advance the prosperity and quality of life of their citizens. Presented biennially, the award encourages innovation in public policy, projects, business models and practices.

The Guangzhou Institute for Urban Innovation ("GIUI") was established in 2012 under the framework of the Guangzhou Award. The Institute is conceived as an international network of experts and institutions dedicated to furthering urban innovation concepts, tools and methodologies. Its members include research and academic institutions, professional associations, community-based organizations, industry leaders and individual experts.

Mission

1. Promoting the evaluation and research compiled by the Guangzhou Award.
2. Exploring and constructing a complete theoretical framework for international urban innovation.
3. Promoting the innovative development and common progress of cities around the world.
4. Offering intellectual support for global urban innovation.

Information sharing and knowledge management

The urban innovation database serves as a knowledge bank of peer reviewed initiatives in urban innovation and analytical case studies and reports. It is regularly updated with each cycle of the Guangzhou Award for Urban Innovation and currently includes more than 400 initiatives from 60 countries. The database is interactive allowing users to carry out searches using several criteria. Users are also invited to submit their stories, articles, views and reviews of examples of urban innovation.

In addition to this knowledge base, the Guangzhou Institute disseminates recurrent and periodic publications and newsletters dedicated to furthering lessons learned and knowledge gained from innovative urban initiatives. Recurrent publications include write ups of 15 shortlisted initiatives from each Award cycle and reports on lessons learned from field visits and study tours to the same initiatives as well as others. Ad hoc publications are also produced by the partners of the Institute and by independent experts.

Office of the Guangzhou Institute for Urban Innovation, 2016.

Box 1.2—5 Cities Win the 6th Guangzhou International Award for Urban Innovation

On the night of December 7, 2023, the award ceremony of the 6th Guangzhou International Award for Urban Innovation (Guangzhou Award) was held at Guangzhou Yuexiu International Congress Center. **Bogotá, Colombia; Chalandri, Greece; Gwangju, Republic of Korea; Kampala, Uganda; and Xianning, China** won the 6th Guangzhou Award. Among them, Xianning, China also claimed the honor of City of Your Choice.

At the ceremony, short films about Guangzhou Award and the 15 shortlisted initiatives were presented and the Chair of the 6th Guangzhou Award Jury announced the selection results. Ten cities, including Antalya, Türkiye; Cape Town, South Africa; Iztapalapa, Mexico; Jakarta, Indonesia; Kazan, Russia; Mannheim, Germany; Pimpri Chinchwad, India; Ramallah, Palestine; São Paulo, Brazil; and Tehran, Iran received Medals of Honor of the Guangzhou Award. Subsequently, representatives from Bogotá, Chalandri, Gwangju, Kampala, and Xianning, the five winners of the 6th Guangzhou Award, accepted the award and delivered their acceptance speeches.

The attendees to the ceremony unanimously agreed that Guangzhou Award has created a new, global platform for cities to exchange urban innovation experience, deepened the city-to-city learning on urban innovation, sparked new changes and fresh insights in urban innovation, and in doing so, propelled the urban innovation and sustainable development of cities worldwide.

Over the past 11 years, six cycles of Guangzhou Award have been held biennially, attracting 1635 initiatives from 653 cities and regions of 101 countries. These initiatives cover a wide range of areas, including infrastructure and public services, urban planning and good governance, partnerships, technology, resilience, social inclusion and gender equality. The Guangzhou Award has now evolved into an important public product that promotes exchanges and city-to-city learning in global urban governance innovation, and a distinguished emblem for Guangzhou to showcase among global cities.

It is learned that the 6th Guangzhou Award received a wide range of responses from cities and regions around the world. A total of 274 initiatives were submitted by 193 cities and regions representing 54 countries and territories, the number of participating cities and initiatives both surpassing those of previous cycles.

Source: Guangzhou Foreign Affairs Office.
(Available online at: http://www.guangzhouaward.org/a/3469.html?lang=en).

Box 1.3—the 6th Guangzhou Award-Collection of the 30 Deserving Initiatives

The biennial Guangzhou International Award for Urban Innovation (Guangzhou Award) has been co-hosted by the United Cities and Local Governments (UCLG), the World Association of the Major Metropolises (Metropolis) and the City of Guangzhou since 2012. Besides recognizing local government excellence, the Guangzhou Award is designed as a learning and exchange platform to help accelerate the local implementation of the global agendas.

This compilation of case profiles is derived from the 6th cycle of the Guangzhou Award, during which 274 initiatives from 193 cities and regions from 54 countries made successful submissions. All of them were assessed by an independent Technical Committee of 11 experts in September 2023. The criteria used included the social, economic, environmental, and political context of each case and their innovativeness, effectiveness and potential for going to scale. As a result, of these 274 initiatives, 45 were identified as exemplary initiatives. This collection presents 30 of the 45 exemplary initiatives as a supplement to the shortlisted 15 outstanding initiatives.

The 30 initiatives address all 17 of the SDGs with most of the initiatives covering several goals and targets simultaneously. They provide an excellent demonstration of the localization of the global agendas. They furthermore provide useful insights as to how many of the processes can be replicated and brought to scale across borders and jurisdictions.

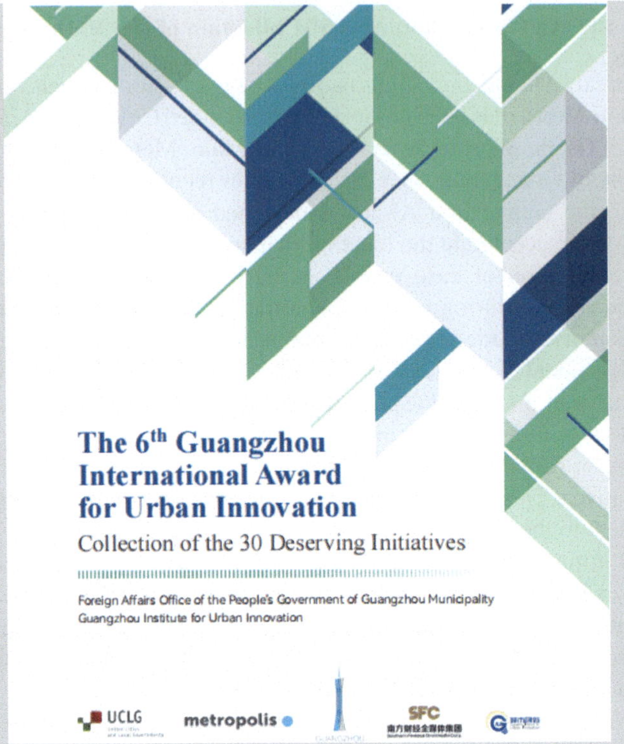

Box 1.4—2023 Project Categories

2023 project entries are in six categories of:

- Social
- Economic
- Environment
- Governance/Management
- Technology
- Others

Database available from: http://www.urban-innovations.org/

Box 1.5—Summary of All Listed Cities

The entries include winning cities (5 in total), the city of your choice (1 city), shortlisted cities (10 as part of 15 cities), and deserving cities (30 in total).

Available from:
http://www.guangzhouaward.org/c/e_2023.html

Winning Cities
Bogotá, Colombia
Halandri, Greece
Gwangju, Republic of Korea
Kampala, Uganda
Xianning, China
The City of Your Choice
Xianning, China
Shortlisted Cities
Tehran, Iran
São Paulo, Brazil
Ramallah, Palestine
Pimpri Chinchwad, India
Mannheim, Germany
Kazan, Russia
Jakarta, Indonesia
Iztapalapa, Mexico
Cape Town, South Africa
Antalya, Türkiye
Deserving Cities
Tlajomulco de Zúñiga, Mexico
Tharaka Nithi, Kenya
Thane, India
Shanghai, China
Salvador, Brazil
São Paulo, Brazil
Ramallah, Palestine
Rotterdam, Netherlands
Recife, Brazil
Quezon City, Philippines
Palmira, Colombia
Nilüfer, Türkiye
Moscow, Russia
Medellín, Colombia
Konya, Türkiye
Krasnoyarsk, Russia
Istanbul, Türkiye(1)
Istanbul, Türkiye(2)
Gwanak-gu, Seoul, the Republic of Korea
Gimhae-City, the Republic of Korea
Guangzhou, China
eThekwini, South Africa
Dubai, United Arab Emirates
Daggar, Pakistan
Barcelona, Spain
Baia Mare, Romania
Bontang City, Indonesia
Berlin, Germany
Buenos Aires, Argentina(1)
Buenos Aires, Argentina(2)

Box 1.6—Report of the Technical Committee

The Technical Committee (TC) met in Guangzhou 10 to 14 September 2023 to select and shortlist outstanding initiatives with a view to recognize and reward the local implementation of Sustainable Development Goals (SDGs) and the New Urban Agenda (NUA). It took into consideration the goal of the Guangzhou International Award for Urban Innovation (Guangzhou Award) to:

- Highlight exemplary models of innovative policies and practices in implementing the urban dimension of the SDGs and the NUA by cities and regions.
- Motivate cities and regions to further invest in and upscale innovation and to learn from each other.
- Improve urban governance through demonstrating how innovation can provide a better urban future for all.

Full report is available online at: http://www.guangzhouaward.org/a/3387.html?lang=en

Box 1.7—Jury Report Summary

The results are support by a comprehensive jury report, which highlights views and suggestions for each initiative or project.

The document is available online at: http://www.guangzhouaward.org/a/3503.html?lang=en

Box 1.8—Urban Innovation and the Implementation of the New Urban Agenda

(Available from: http://www.guangzhouaward.org/a/3056.html?lang=en).

This report is consisted of three chapters around the three transformative commitments highlighted in the 2022 Progress Report on the Implementation of the New Urban Agenda (NUA): Social Inclusion and Ending Poverty, Inclusive Urban Prosperity and Opportunities, and Environmental Sustainability and Resilience. Each chapter selects shortlisted and award-winning finalists from the last two cycles of the Guangzhou Award (2018, 2020), covering the geographical regions of Africa, West Asia, North Africa, Asia Pacific, Europe, Latin America and the Caribbean, and North America. These initiatives have been selected to illustrate how local governments are turning ideas into action through practical and institutional innovations in governance, planning and financing practice, to achieve urban inclusion and resilience.

The report highlights that, regarding social inclusion and ending poverty, citizen engagement and participatory governance effectively contribute to community resilience and social inclusion. Regarding inclusive urban prosperity and opportunity, transformative economic models start with cities and regions, where decent work and inclusive urban growth could be promoted by circular economy and multi-sectoral cooperation. In terms of environmental sustainability and resilience, bottom-up and top-down inter-governmental dialogue and local capacity building are required to achieve desired outcomes and combining environmental sustainability with measurable benefits is an important step in driving sustainable change.

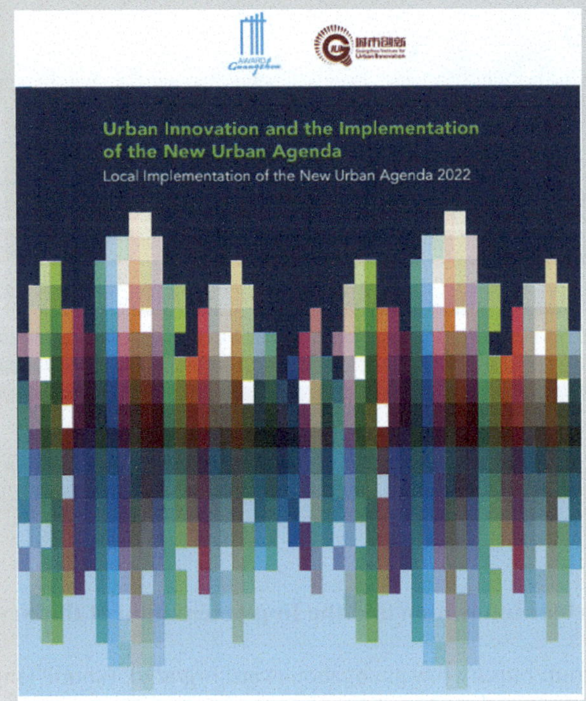

Box 1.9-a Collection of Shortlisted Initiatives of the 6th Guangzhou Award
(Available from: http://www.guangzhouaward.org/a/3455.html?lang=en).

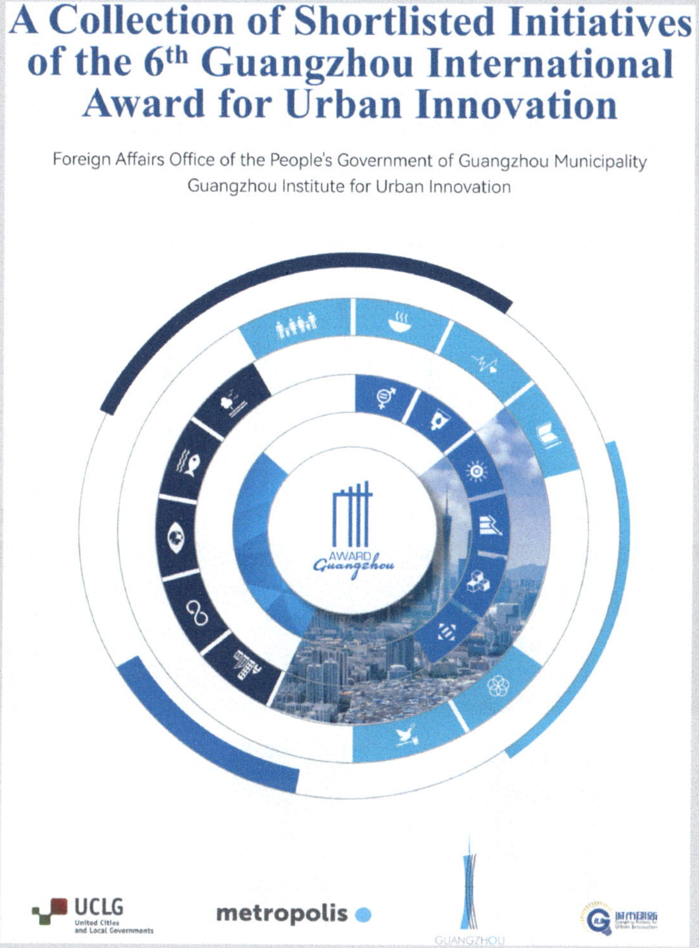

Box 1.10—Guangzhou Award Urban Innovation Database

The International Database on Urban Innovation contains all the original materials of over 1300 initiatives submitted to the Guangzhou Award in the past ten years. Shortlisted initiatives of each cycle of the Award have been selected into the Urban Best Practices Database of the UN-Habitat. In 2022, the database has been comprehensively upgraded to further promote exchanges and mutual learnings among cities and local governments around the world, and thus provide reference for sustainable global urban development.

Summary of database for the first five cycles and key reports are shown below.

Available from: http://www.guangzhouaward.org/The6thGuangzhouAward?lang=en

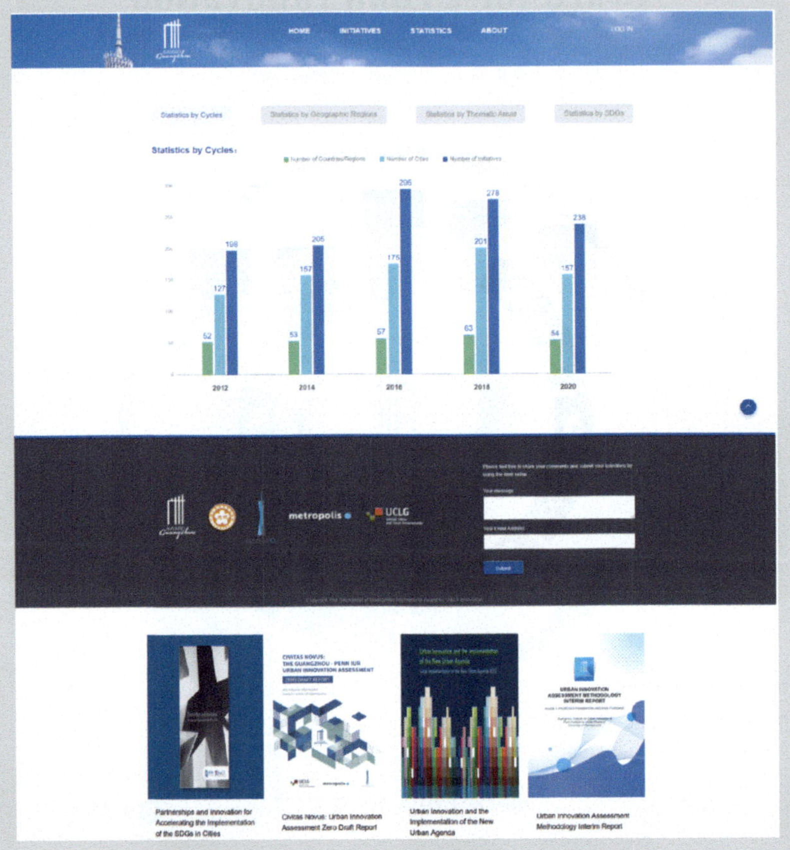

References

1. Alcamo J, Thompson J, Alexander A et al. (2020). Analysing interactions among the sustainable development goals: findings and emerging issues from local and global studies. *Sustainability Science*, 15:1561–1572
2. Allam Z, Cheshmehzangi A (2024). Sustainable futures and green new deals. Singapore and Germany: Palgrave Macmillan

3. Cheshmehzangi A, Dawodu A, Cheshmehzangi A, Dawodu A (2019). The review of sustainable development goals (SDGs): People, perspective and planning. Sustainable urban development in the age of climate change: People: The cure or curse, Singapore and Germany: Palgrave Macmillan. pp. 133–156
4. Cheshmehzangi A, You N, Siri J, Birch E (2024) Harnessing urban innovation to unlock the sustainable development goals. Singapore: Springer.
5. Cheshmehzangi A, Zou T (2024). Literary analysis of the connection between urban innovations and sustainable development goals (SDGs). Sustainability. 16(15):6405
6. Dalby S, Horton S, Mahon, R, Thomaz D (2019). Achieving the sustainable development goals. Global governance challenges. New York: Routledge Taylor and Francis
7. Dawodu A, Sharifi, A, Cheshmehzangi A, Oladejo, J (2021). The illusion of participation: Are participatory indicators truly effective in neigborhood sustainability assessment tools. J. Clean. Prod. 311:127538
8. Grainger-Brown J, Malekpour S, Raven R, Taylor E (2022) Exploring urban transformation to inform the implementation of the sustainable development goals. Cities. 131:103928
9. Lotfata A (2023) Urban sustainability: Multifunctional and multipurpose planning of urban space. In: The Palgrave Encyclopedia of Urban and Regional Futures. Cham: Springer International Publishing. pp. 2087–2092
10. Moussa RA (2023) A responsive approach for designing shared urban spaces in tourist villages. Sustainability 15(9):7549
11. United Nations (2021). 2021 Sustainable development forum People-centered sustainable development. Available from: https://www.un.org/en/desa/2021-sustainable-development-forum-people-centered-sustainable-development

Chapter 2
Eco-Friendly Farmer Card: Protect Your Environment, Collect Points, and Win Prizes in Antalya, Türkiye

Abstract The municipality of Antalya has initiated the innovative Farmer Card Project, aimed at addressing the urgent issue of hazardous agricultural packaging waste, which is a significant environmental and health concern in the city due to intense farming activities. The main goal of the project is to reduce plastic pollution from pesticide-laden packaging, which negatively impacts soil, causes pollution, and indirectly pollutes 98% of Antalya's drinking water. With the large scale of agricultural activities in the area, waste management has become an essential priority.

This groundbreaking initiative introduces a sustainable system that includes solar-powered, fire-resistant smart containers called "Waste Vending Machines." These machines are strategically positioned in agricultural areas and are monitored 24/7, using magnetic keyfobs provided to farmers for access. When farmers deposit agricultural waste packaging, it is automatically weighed, and they earn reward points on their keyfobs. These points can be redeemed for agricultural supplies or discounts, encouraging responsible disposal practices. An associated mobile app, "Çevre Dostu Çiftçi" (Eco-Friendly Farmer), enables farmers to keep track of their points, rewards, and find the nearest vending machines, making the process convenient and accessible. This project not only reduces environmental damage but also promotes a culture of sustainability among local farmers, motivating them to embrace greener practices. The initiative is related to SDGs 2, 6, 7, 9, 12, 13, 14, 15, and 17.

Keywords Eco-friendly farmer card project · Agricultural waste management · Plastic pollution · Sustainable farming practices · Waste vending machines · Environmental innovation · Pesticide packaging recycling

2.1 Introduction to Antalya, Türkiye

Antalya, positioned on Turkey's southwestern coast, is a lively city steeped in history and is one of the country's most important urban areas. Situated along the Mediterranean Sea, Antalya is famous for its magnificent natural landscapes, cultural legacy, and economic vibrancy. Boasting a population of 2,735,791, the city serves as a major

A. Cheshmehzangi et al., *People-Centered Innovation*, Urban Sustainability,
https://doi.org/10.1007/978-981-96-5230-3_2

center for tourism, agriculture, commerce, and other vital industries, driving its prosperity. With its population exceeding 2.7 million, Antalya is a thriving and expanding urban region. The city exhibits a population growth rate of 2.60%, reflecting its ongoing progress and allure as a desirable place to reside, work, and explore. This development is fueled by various factors, such as a robust economy, favorable climate, and high quality of life. The migration of both local and international individuals seeking opportunities in tourism, agriculture, and other fields has further contributed to this population surge.

Covering an area of 20,909 square kilometers, Antalya ranks among the largest provinces in Turkey. Despite its expansive land mass, the city's population density remains relatively low at 129 individuals per square kilometer. This low density is evidence of the city's well-balanced development, which has preserved extensive natural landscapes, including mountains, forests, and coastlines. Antalya's geographical variety not only enhances its appeal as a tourist hotspot, but also supports its flourishing agricultural sector, benefiting from fertile lands and a favorable Mediterranean climate.

Antalya enjoys a diverse economy, benefiting from various sectors. It enjoys a GDP per capita of $10,665, signifying a comparatively high quality of life in comparison to other Turkish regions. The city's economic vitality stems from its primary sources of prosperity: tourism, agriculture and livestock, commerce, mining, and health.

Antalya's economy thrives on tourism, drawing millions of visitors annually with its beautiful beaches, historic sites, and rich cultural scene. Known as the "Turkish Riviera," the city's tourism industry is a significant source of employment and income for locals, consistently growing thanks to infrastructure improvements and service investments that enhance its allure to international tourists. Antalya's tourism sector caters to a diverse range of preferences, offering luxury resorts, golf courses, ecotourism, and cultural experiences.

Agriculture is a crucial part of Antalya's economy, as the region is one of Turkey's leading agricultural centers. Its fertile plains are perfect for growing a variety of crops, including fruits, vegetables, and flowers, with a particular emphasis on citrus fruits, olives, and greenhouse-grown vegetables. Livestock farming such as cattle, sheep, and poultry also contributes significantly to the agricultural sector, maintaining Antalya's status as a top agricultural producer in Turkey. Commerce is vital to Antalya's economy, as the city's strategic location and well-developed infrastructure have made it a key commercial hub in the region. Antalya's port facilities and proximity to international markets facilitate the export of agricultural products, textiles, and other goods, while its vibrant retail and service sectors contribute to the local economy.

Though less prominent than tourism and agriculture, mining also contributes to Antalya's economic landscape. The region's natural resources, including marble, chrome, and lignite, provide employment and contribute to the city's industrial output. Despite environmental sustainability challenges, the mining sector remains a significant source of revenue for the local economy. Antalya has established itself as a hub for health tourism, attracting patients from across the globe seeking medical treatments and wellness services. The city's modern hospitals, skilled healthcare professionals, and competitive pricing make it a sought-after destination for medical tourism. As Antalya continues to invest in healthcare infrastructure and services, the health tourism sector is projected to expand further, offering a wide range of treatments from cosmetic surgery to advanced medical procedures.

Antalya, despite its economic prosperity, grapples with issues of social and economic disparity. With a GINI Index of 0.4, the city exhibits a moderate level of income inequality. While a significant portion of the population benefits from the thriving economy, others, particularly those in rural areas or employed in low-paying sectors, may not have the same access to economic opportunities. It is crucial to address these discrepancies to ensure that the advantages of growth are distributed more fairly across all segments of the community.

To summarize, Antalya is a city characterized by both contrasts and opportunities, where contemporary innovation meets ancient history, and a thriving economy coexists with natural beauty. Its increasing population, varied economy, and strategic location position it as a significant player in Turkey's development. Nonetheless, as the city continues to expand, it must also confront the challenges associated with inequality and sustainable development to guarantee a prosperous future for all its inhabitants.

2.2 Background and Origins

Antalya is an important hub for agriculture, with a focus on greenhouse farming to meet the demand for fruits and vegetables both locally and internationally. The region is renowned for its exceptional productivity, particularly in the cultivation of tomatoes, cucumbers, peppers, and citrus fruits, which are crucial for local markets and exports. However, the significant expansion of greenhouse farming has resulted in a proportionate increase in pesticide usage, with Antalya accounting for 10.1% of Turkey's annual 50,000-ton pesticide consumption. The extensive use of pesticides, essential for maintaining high crop yields, poses substantial environmental risks by causing habitat degradation, soil and water contamination, and a negative impact on biodiversity. These chemicals can infiltrate groundwater, affecting 98% of Antalya's drinking water sources, and contribute to long-term soil depletion, reducing its fertility over time.

The primary goal of the project is to develop innovative solutions to address these urgent issues and secure the sustainability of Antalya's agricultural sector. By focusing on responsible management of pesticide packaging waste and promoting environmentally friendly practices, the project aims to preserve the fertile agricultural land and critical water resources of the region for future generations. This initiative not only mitigates environmental harm but also supports the long-term viability of agriculture in Antalya, ensuring its role as a cornerstone of the local economy.

2.2.1 Key Innovative Aspects

In the agriculturally intensive areas of Finike and Kumluca, a significant environmental concern arose due to excessive agricultural packaging waste, as identified by experts. These districts are recognized for their extensive greenhouse farming, leading to the generation of large quantities of pesticide containers and other hazardous packaging materials. The Departments of Environmental Protection, Zero Waste, and Agriculture collaborated on a comprehensive solution after realizing the urgency of the issue. This solution was formulated based on in-depth interviews with farmers to gain insights into the waste management challenges specific to the agricultural community.

The project was further enhanced by the involvement of the private sector, which introduced innovative equipment. This included user-friendly key-chain cards for farmers to use in accessing smart waste vending machines. The software, accessible on smartphones and computers, allows for real-time data monitoring and scoring based on the 1 g = 1 point criterion, incentivizing farmers to actively participate in the waste disposal program.

Farmers participating in the program are offered valuable incentives, such as air-heat-humidity sensors, greenhouse materials, and essential farming tools. Furthermore, they receive training sessions and guidance on identifying hazardous packaging waste and properly disposing of it in the designated vending machines. This exclusive service, available only to registered farmers, significantly contributes to environmental sustainability while promoting a circular economy and reducing carbon emissions. This initiative marks a significant advancement in Turkey's environmental efforts and sets a precedent for future waste management strategies in the agricultural sector. The project's aim is to create a cleaner, safer, and more sustainable environment for future generations by effectively addressing the waste problem in Finike and Kumluca.

2.2.2 Desired Outcome and Measurement

2.2.2.1 Desired Outcome

The project's goal is to increase environmental awareness among Turkish farmers by adopting the circular economy model, moving away from the punitive "The Polluter Pays" approach to a more positive "The Protector Wins" strategy. This new model encourages farmers to adopt sustainable practices and rewards them for their environmental stewardship. Farmers can earn prizes by accumulating points through the responsible disposal of pesticide packaging waste, which not only protects public health but also creates a strong incentive for sustainable behavior.

The business model is based on the take-make-dispose framework but has been modified to promote recycling and waste reduction. The establishment of the Zero

Waste Branch Directorate has significantly improved the institutional capacity to manage and expand the project. The project was initiated in 2016 in Antalya, Turkey, a region that ranks third globally in pesticide packaging waste and has made significant progress in changing waste management practices.

In 2022, the project was chosen as a pilot program and introduced smart containers at six key locations in areas with high agricultural production. These smart containers have simplified farmer participation, contributing to the project's success. As the project continues to be implemented sustainably, it is also expanding in scale, with the aim of reaching more districts and further integrating sustainable practices in Turkey's agricultural sector.

2.2.2.2 Measurement

The project systematically collects detailed information from participating farmers to enhance the effectiveness of the waste management system and tailor incentives accordingly. This data includes essential details such as the farmer's name, farmer number, field area, crop specifics, and the amount of agricultural packaging waste they generate. Additionally, the project gathers information about the irrigation status of the farms, which is crucial for understanding the broader environmental impact of farming practices and for fine-tuning the support provided to the farmers.

A key aspect of the project is tracking the quantity of agricultural packaging waste deposited into the smart vending machines using the designated cards. This waste is automatically measured by sophisticated software that links directly to the reward system, ensuring that farmers are accurately rewarded for their contributions to environmental sustainability. The integration of such technology not only streamlines the process but also creates a transparent and accountable system that encourages participation.

To maintain trust and encourage widespread involvement, the project strictly adheres to personal data protection regulations in the country. Personal information such as age, gender, income group, and any disabilities of the farmers is kept confidential, ensuring privacy and security. This commitment to data protection reassures participants that their personal information is handled with the utmost care, fostering a sense of security and willingness to engage with the project. By safeguarding this sensitive information while simultaneously collecting relevant agricultural data, the project effectively balances innovation with respect for individual privacy, setting a new standard for responsible data use in environmental initiatives.

2.2.3 *Implementation*

2.2.3.1 Principal Parties and Partners

The waste management program demonstrates a comprehensive strategy for addressing agricultural packaging waste by promoting collaboration across multiple sectors. Fundamental to this program is the cooperation between metropolitan and district municipalities, which ensures efficient waste collection and storage. Citizens also play a vital role in the proper disposal of waste, forging a strong partnership between the public and citizens that expands the program's reach and effectiveness.

An essential component of the program is the collaboration with a local company that manufactures innovative waste vending machines. These machines are designed with sustainability in mind, featuring solar panels that enable them to operate independently for up to 60 days. They also incorporate advanced features such as impact sensors, cameras, GPS tracking, and fire prevention mechanisms to ensure functionality, security, and reliability.

Furthermore, the initiative involves academia, which provides research and insights to continuously enhance the system. This unique collaboration addresses agricultural packaging waste holistically, utilizing innovative coating techniques to prevent environmental hazards like leachate. The technologies and methods developed have the potential to be adapted and implemented in other countries with similar agricultural practices, establishing this initiative as a global model for agricultural waste management.

Looking to the future, the program aims to focus on software improvements and capacity enhancements, which are expected to drive economic development. By increasing the number of suppliers, service providers, and employment opportunities, the initiative seeks to strengthen the capabilities of companies in various cities and municipalities. This expansion not only promises environmental benefits but also significant economic growth, showcasing the program's potential as a catalyst for sustainable development in the region and beyond.

2.2.3.2 Resources for Implementation

The use of municipal funds is crucial for advancing the principles of the circular economy, which prioritize sustainable resource utilization and waste minimization. The project aims to establish a closed-loop system, emphasizing the continual reutilization, recycling, and repurposing of materials. This endeavor demonstrates these principles by focusing on material recycling, energy production from waste, development of alternative materials, sustainable production methods, and promotion of innovation.

Operating within the circular economy framework, the project seeks to shift from the traditional linear model of take-make-dispose to one that eliminates waste and pollution. By maximizing the utilization of products and materials, the initiative endeavors to decrease the environmental impact of agricultural activities. Additionally, it contributes significantly to creating job opportunities and driving economic growth within the community.

An essential part of the project's financial strategy is its endeavor to transform people's attitudes toward waste management. By moving away from the conventional "Polluter Pays" approach to the more proactive "Protector Wins" philosophy, the initiative challenges long-held misconceptions about waste responsibility. This change in perspective encourages individuals to view themselves as contributors to environmental sustainability, rather than just bearing the cost of pollution. This shift in mindset promotes a more comprehensive and effective method of waste management, aligning with the broader objectives of the circular economy and promoting a culture of environmental stewardship.

2.2.3.3 Tools and Methods Applied

In the past, farmers in the area frequently used harmful methods like burning to get rid of agricultural waste, leading to air pollution and other environmental issues. The initiative tackles these challenges by promoting clean energy options and encouraging innovation, which improves both economic and environmental sustainability. Shifting away from damaging waste management techniques not only decreases pollution but also aligns with sustainable farming methods.

The project has significantly increased public awareness regarding the need for responsible waste management and environmental protection. It actively involves various stakeholders, including local communities, government bodies, and private sector collaborators, to devise and execute effective solutions. Through collaborative research and development (R&D) efforts in a multidisciplinary setting, the project creates inventive strategies and technologies. These endeavors help establish enduring solutions that benefit the environment and the economy, showcasing a dedication to sustainable practices and establishing a model for future endeavors.

2.3 Obstacles and Solutions

The implementation of the "Eco-Friendly Farmer Card Project" faced several notable obstacles and challenges, each requiring targeted strategies to overcome (Fig. 2.1).

Awareness and Education: One of the primary challenges was the initial lack of awareness among farmers regarding the benefits of the program and the significance of proper waste disposal. Many farmers were unfamiliar with how their participation could positively impact the environment and their own practices. To address this issue, the project introduced comprehensive training programs and a reward system to engage farmers effectively. These educational initiatives aimed to raise awareness, illustrate the benefits of the program, and demonstrate the importance of proper waste management. Interactive workshops, field demonstrations, and informational

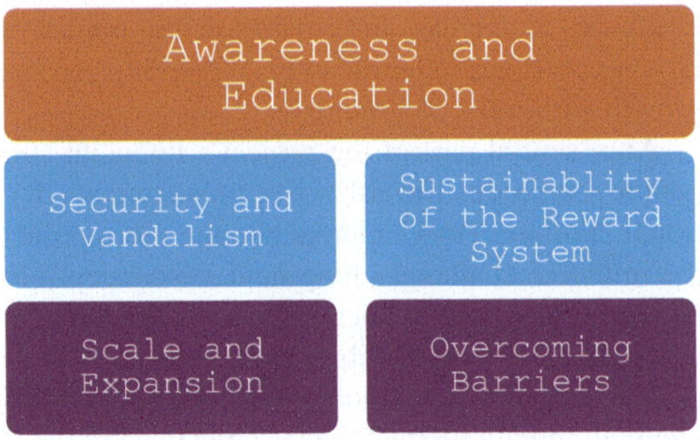

Fig. 2.1 Target strategies of the project

materials were developed to foster a deeper understanding and encourage active participation.

Security and Vandalism: Another challenge was ensuring the security of the vending machines and digital systems. Vending machines, being high-value assets, are susceptible to vandalism or theft. To mitigate these risks, the project equipped the machines with advanced security features, including alarm systems and surveillance cameras. These measures were designed to protect the machines from tampering and theft while ensuring the integrity of the data collected. Regular maintenance and monitoring protocols were also established to address any security concerns promptly.

Sustainability of the Reward System: Maintaining the reward system's sustainability over the long term proved challenging. Ensuring a consistent and reliable source of funding or incentives to motivate farmers was a significant barrier. The project team explored various funding models, including public–private partnerships and sponsorships, to secure ongoing financial support. Additionally, efforts were made to diversify the types of rewards and incentives, making them more appealing and relevant to farmers.

Scale and Expansion: Scaling up the project presented logistical, financial, and administrative challenges. Expanding the initiative to cover more regions required careful planning and resource allocation. The project team had to address the complexities of scaling operations, including coordinating with additional stakeholders, managing increased waste volumes, and maintaining consistent service quality across new areas. Significant investment in infrastructure, technology, and personnel was necessary to support the expansion.

Overcoming Barriers: Addressing these barriers effectively required a multifaceted approach. Strong government support was essential to provide the necessary resources and policy backing. Community engagement played a crucial role in ensuring local buy-in and participation. Continued education efforts were needed to keep farmers informed and motivated. An ongoing commitment to sustainability and environmental protection was vital to the project's long-term success. By combining these strategies, the project aimed to overcome its challenges and create a lasting impact on waste management practices and environmental sustainability.

2.3.1 Achieved Outcome

Throughout the implementation of the Eco-Friendly Farmer Card Project, significant strides were taken to address agricultural packaging waste and advocate for sustainable practices. In Kumluca and Finike districts, the project strategically installed 13 state-of-the-art waste vending machines, evenly distributing 6 machines in each area. These machines engaged a total of 2041 farmers, offering an effective and creative solution for handling agricultural waste. During the course of the project, these vending machines amassed 1200 kg of waste, signifying a remarkable decrease in environmental pollution. The collected waste was efficiently processed and recycled, reintegrating into the circular economy and diminishing the ecological impact of agricultural activities. As a token of appreciation, 21 users received rewards, with 17 farmers being acknowledged in the initial phase. The rewards, comprising of agricultural tools and equipment, acted as incentives to promote more sustainable practices and instigate a shift in behavior among farmers.

The project also played a pivotal role in cultivating environmental consciousness among the participating farmers. By educating them about the advantages of proper waste disposal and its impact on sustainability, the project effectively advocated for more responsible and environmentally friendly farming practices. To ensure openness and answerability, the municipality disseminated waste data and project outcomes through various channels, including public ceremonies, reports from agricultural chambers, annual reports, and comprehensive inventory data. These measures not only highlighted the accomplishments of the project but also fostered trust and involvement within the broader community.

Looking to the future, the triumph of the project in Antalya has set the stage for plans to extend the initiative nationwide. This expansion aims to advance broader environmental sustainability throughout Turkey, leveraging the insights gained and successes achieved to enhance waste management practices on a larger scale.

2.4 Lessons Learned

Antalya's initiative to build a localized recycling and disposal network through the Eco-Friendly Farmer Card represents a significant advancement in waste management and sustainability. By leveraging this innovative card system, the municipality can address various challenges associated with agricultural packaging waste, even in the face of external pressures such as economic fluctuations or logistical constraints. The card incentivizes farmers to adopt recyclable and eco-friendly packaging, thereby promoting a circular economy that prioritizes waste reduction and resource efficiency.

The introduction of the eco-friendly farmer card has several key advantages. It encourages farmers to engage in sustainable practices by providing tangible rewards for their participation, thus fostering a culture of environmental responsibility. Additionally, the system supports real-time data collection and analysis, allowing Antalya to closely monitor waste generation, recycling efficiency, and farmer engagement. This data-driven approach enables evidence-based improvements to the program, ensuring that it remains effective and responsive to evolving needs.

The success of this initiative offers valuable lessons in localized waste management. It highlights the importance of integrating health measures with environmental practices, engaging the community through educational efforts, and adopting technology to streamline processes. Adhering to circular economy principles, fostering collaboration among stakeholders, and practicing adaptive planning are also crucial elements that contribute to the program's success.

Furthermore, the program demonstrates a long-term commitment to sustainability and environmental protection, aligning with global efforts to create more sustainable and environmentally responsible communities. By addressing critical challenges such as waste management and supply chain resilience, the initiative not only improves local environmental conditions but also sets a precedent for other regions seeking to implement similar solutions. This proactive approach reflects a comprehensive understanding of the interconnected nature of environmental, economic, and social factors, paving the way for future advancements in sustainable waste management practices.

Related SDGs: Goals 2, 6, 7, 9, 12, 13, 14, 15, and 17.

Acknowledgements We explicitly thank the project owners based in Antalya, Türkiye for their participation and the provision of materials for the 6th Guangzhou International Award for Urban Innovation. We also congratulate them for their achievements in the 2023 cycle of awards.

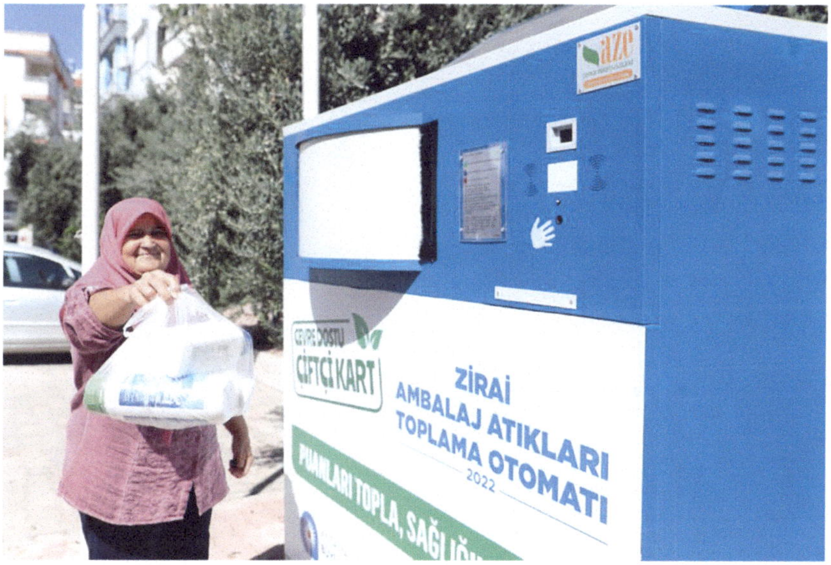

The above images are provided to us by the project owners for their submission.

Chapter 3
Bogotá Care Blocks: How a City is Reorganizing Itself for Women and Achieving the SDGs in Bogotá, Colombia

Abstract Women bear the majority of the unpaid care burden, resulting in various negative consequences such as limited financial independence, higher rates of dropping out of school, persistent mental and physical health challenges, and heightened susceptibility to gender-based violence. Bogotá has taken the lead in Latin America by establishing the first city-level Care System to tackle these obstacles. This innovative system, centered around Care Blocks, works to acknowledge, diminish, and redistribute care responsibilities. Care Blocks employ a unique "ease-of-access" approach that has not been tested on this scale before, offering educational opportunities, leisure activities, and income-generating options for caregivers. Additionally, they provide professional care for dependents, including children, individuals with disabilities, and the elderly. The initiative also includes a cultural shift strategy, seen in the creation of a "Care School for Men" aimed at educating men about recognizing and equitably sharing care responsibilities within households. As a form of social innovation, Care Blocks are intended to combat economic and time-related poverty, champion gender equality, and promote environmental sustainability. This model has already proven its effectiveness and is easily reproducible, making it a promising framework for other cities addressing similar issues. The initiative is related to SDGs 3, 5, 11, 16, and 17.

Keywords Unpaid care burden · Gender equality · Care blocks · Economic autonomy · Social innovation · Cultural change strategy · City-level care system

3.1 Introduction to Bogotá, Colombia

Bogotá, Colombia's capital city, is a lively metropolis situated in the heart of the nation, both geographically and culturally. With a population of around 7,834,167, Bogotá is a dynamic urban center that plays a key role in Colombia's political, economic, and social landscape. The city's population growth rate of 1.24% demonstrates its continued development and the increasing urbanization trends in the country. Encompassing an area of 1776 square kilometers, Bogotá is known for its

A. Cheshmehzangi et al., *People-Centered Innovation*, Urban Sustainability,
https://doi.org/10.1007/978-981-96-5230-3_3

high population density of 21,276 people per square kilometer. This density signifies the city's bustling nature and its function as a hub for various activities and services. As Colombia's largest city, Bogotá is a vital component of the nation's infrastructure and a focal point for economic and cultural exchanges.

Bogotá's economic landscape is characterized by a GDP per capita of $12,117 as of 2022, positioning it as a significant economic powerhouse within Colombia. The city's strong service sector, including finance, education, health, and technology, is the primary source of prosperity. This diverse service economy underpins Bogotá's status as an important business and financial center in Latin America. However, Bogotá faces significant challenges, as evidenced by its GINI Index of 0.53. This measure of income inequality underscores the disparities within the city, emphasizing the need for targeted policies and initiatives to address socio-economic imbalances and enhance inclusivity.

Apart from its economic importance, Bogotá possesses a rich cultural and historical heritage. It is home to numerous museums, historical sites, and cultural institutions that reflect its colonial past and its role in shaping Colombian identity. The city's cultural landscape is further enriched by its diverse population, contributing to a vibrant tapestry of traditions, languages, and practices. Bogotá's urban structure is continuously evolving, with ongoing efforts to tackle various challenges related to infrastructure, transportation, and housing. The city has undertaken several ambitious projects aimed at improving urban mobility, reducing traffic congestion, and enhancing the quality of life for its residents. Initiatives such as the expansion of public transportation networks and the development of green spaces are central to these efforts, aiming to create a more sustainable and livable urban environment.

Bogotá has made significant progress in tackling social issues and promoting inclusive growth in recent years. An example of this is the city's innovative approach to addressing the unequal burden of unpaid care, which particularly impacts women. The introduction of the Care System at the city level through Care Blocks is a pioneering effort to acknowledge, diminish, and distribute care responsibilities. This initiative not only aims to ease the challenges faced by caregivers but also aims to promote greater gender equality and social unity. Bogotá's transformation into a prominent Latin American city is characterized by its dynamic expansion, economic vigor, and continuous endeavors to tackle social and environmental obstacles. The city's dedication to innovation and sustainability, coupled with its diverse population and rich cultural heritage, establishes it as a leading model of urban development in the twenty-first century. As Bogotá continues to navigate the complexities of modern urban life, its initiatives and projects will have a crucial role in shaping its future and enhancing its position as a key player in the global urban landscape.

3.2 Background and Origins

The development and implementation of Care Blocks is a key priority in Bogotá's four-year governance plan led by Claudia López-Hernández, the city's first female mayor. The current administration has made significant progress in addressing the fundamental challenges faced by women in the city, particularly those associated with the unpaid care responsibilities. Care Blocks are now an essential element of the Master Urban Plan, which outlines Bogotá's development path until 2035. The plan aims to establish 45 Care Blocks, demonstrating a long-term commitment to creating a more balanced and supportive urban environment.

In addition to their inclusion in the Master Urban Plan, Care Blocks have been officially integrated into city legislation by the City Council, ensuring their continuity beyond the current administration. This legal support guarantees that the Care Blocks initiative is not just a temporary measure but a sustained initiative to tackle the systemic issues related to care work and gender inequality in Bogotá.

The "time poverty" experienced by women is a significant structural factor contributing to gender inequality in the city. In Bogotá, the disproportionate impact of the unpaid care responsibilities on women is evident in alarming statistics: 30% of women are engaged in full-time unpaid caregiving, with 91% coming from low-income backgrounds and 33% lacking free time for personal care. This underscores the urgent need for systematic changes to ease the burdens on women and provide them with the necessary support. Over time, women's groups in Bogotá have advocated for a comprehensive care system that recognizes and addresses the demands of care work. These organizations have consistently pushed for reforms to reduce, redistribute, and acknowledge care responsibilities. In response to these calls, Claudia López-Hernández pledged during her 2019 mayoral campaign to establish a care system if elected. In line with her promise, her administration introduced the Bogotá Care System in 2020, featuring Care Blocks and their mobile version, Care Buses.

Civil society involvement has profoundly influenced the design and implementation of Care Blocks. Women's organizations, community groups, and other stakeholders have played a critical role in shaping the system to ensure it meets the actual needs of caregivers and those under their care. This participatory approach has been vital in tailoring the Care Blocks to offer essential services and support, including educational resources, leisure activities, and income-generating opportunities for caregivers, alongside professional care for children, the elderly, and individuals with disabilities. By taking an inclusive and creative approach, Bogotá is meeting the immediate needs of its residents while also establishing an example of how cities can tackle care work and gender equality. The Care Blocks initiative is a pioneering effort to reshape the city's social infrastructure and promote a fair and supportive urban environment for everyone.

3.2.1 Key Innovative Aspects

The concept of Care Blocks in Bogotá marks a significant advancement in urban care systems, characterized by key innovations that set them apart from traditional methods. At the core of this innovation lies the centralization of caregivers. Care Blocks not only acknowledge the crucial role of caregivers but actively support them, thereby enhancing their quality of life and, consequently, the standard of care they provide. This caregiver-centric model guarantees that those responsible for the essential task of caring for others receive the necessary support and acknowledgment. Additionally, Care Blocks represent a shift in urban planning, reorganizing the city to meet people's needs rather than the opposite. This approach entails a reconsideration of urban space, ensuring that it is designed with the practical aspects of caregiving and care receiving in mind. This change is vital for establishing a system that is responsive and adaptable to the needs of its users.

Another innovative aspect of Care Blocks is the addressing of care inequalities through a cultural perspective. By incorporating cultural awareness into their design and implementation, Care Blocks seek to encourage long-term, sustainable change. This method not only tackles the immediate needs of caregivers and care receivers but also aims to transform societal attitudes towards care work, fostering a more supportive and fair environment.

Furthermore, Care Blocks offer a social solution to environmental challenges. They integrate eco-friendly practices and sustainable resource management, contributing to broader environmental sustainability objectives. This integration of environmental considerations into the care system underscores the capacity of social innovations to simultaneously address multiple societal challenges. The comprehensive public management model of Care Blocks is another significant innovation. Unlike traditional systems that typically operate independently, Care Blocks consolidate various services within a single space, offering simultaneous support for both caregivers and care recipients. This integrated management approach enhances efficiency and ensures that all aspects of care are addressed in a coordinated manner.

Despite the benefits of Care Blocks, there were institutional challenges during their implementation. One major obstacle was fostering cooperation among the 13 public entities involved in providing services within a Care Block. It was essential to align these diverse entities and ensure they worked together towards common goals. The integration of multiple services required the establishment of new procedures and rules to facilitate collaboration among the various entities. This involved developing protocols that enabled effective coexistence and operational harmony within the shared space of Care Blocks. Managing shared budgets posed another significant challenge. It was necessary to develop a unified budgeting system to ensure budget efficiency across all services. This system was crucial for maintaining financial accountability and optimizing resource allocation.

Redesigning the data collection system was crucial for accurately measuring the impact of Care Blocks. Implementing a new system allowed for comprehensive data gathering and analysis, providing valuable insights into the effectiveness of Care

Blocks and guiding future improvements. This holistic approach to data collection and impact measurement was vital for assessing the success of the initiative and guiding its ongoing development. In essence, Care Blocks in Bogotá exemplify a new and comprehensive approach to urban care systems that address social and environmental challenges through inventive design and thorough management. The initiative showcases the potential for creating sustainable and equitable solutions in urban planning, thereby setting a precedent for other cities facing similar issues. Despite the encountered challenges, the successful implementation of Care Blocks underscores the effectiveness of a coordinated, culturally sensitive, and environmentally conscious approach to care.

3.2.2 Desired Outcome and Measurement

3.2.2.1 Desired Outcome

The primary objectives of Care Blocks in Bogotá are to recognize the crucial role of caregiving and the valuable contribution of caregivers to society, to distribute care responsibilities more fairly among members of society, and to reduce the burden of unpaid care work on women to free up their time for education, formal employment, well-being, and social and political participation. Care Blocks' implementation aims to achieve these objectives through a comprehensive approach that integrates services for both caregivers and care recipients, creating a supportive environment for everyone involved. The initiative prioritizes caregivers and strives to improve their quality of life, which is essential for promoting a more even distribution of caregiving responsibilities in society.

The redistribution of care responsibilities aims to achieve a more equitable division of caregiving tasks across genders and different societal groups. This approach not only seeks to balance the burden of care but also challenges and transforms traditional gender roles and expectations related to caregiving. Reducing women's unpaid care work is a significant focus as it addresses a major obstacle to women's economic and social progress. By easing the excessive burden placed on women, Care Blocks enable them to pursue education, formal employment, and other activities that contribute to their personal growth and societal involvement.

Cultural transformation is a crucial aspect of this initiative, representing a long-term challenge. Bringing about cultural shifts to change entrenched attitudes about caregiving and its value is essential. However, there has been a noticeable shift among caregivers who are starting to recognize their own role and value in society. This evolving self-awareness is critical for establishing a culture that respects and supports caregiving, ultimately leading to more sustainable and equitable caregiving practices. In summary, the Care Blocks initiative is a comprehensive effort to address the multifaceted challenges of caregiving in Bogotá. By focusing on recognition, redistribution, and reduction of care responsibilities, and by fostering cultural change, the program aims to create a more inclusive and supportive environment for caregivers and those they support.

3.2.2.2 Measurement

An impact assessment of Care Blocks and Care Buses is currently underway, employing both quantitative and qualitative methodologies to provide a comprehensive evaluation of the initiative. This assessment aims to measure the effectiveness of the Care Blocks and Care Buses in improving the quality of life for caregivers and those receiving care, as well as the overall impact on the community. Quantitative methods involve the use of structured surveys to gather data on various indicators, such as usage rates of the services, satisfaction levels, and demographic information. These surveys are designed to capture a wide range of data points, including frequency of use, types of services accessed, and perceived benefits. This data is crucial for understanding the scale of the initiative's reach and its effectiveness in meeting the needs of the community.

Qualitative methodologies complement the quantitative approach by providing deeper insights into individual experiences and perceptions. In-depth interviews and focus groups are conducted with a diverse range of participants, including caregivers, care recipients, and community stakeholders. These interviews are carefully designed to explore participants' experiences with Care Blocks and Care Buses, including any challenges faced, perceived improvements, and suggestions for future enhancements. Focus groups are disaggregated by factors such as age, location, type of care, and gender to ensure that the findings reflect the varied experiences of different groups within the community.

By combining these methodologies, the impact assessment aims to provide a nuanced understanding of how Care Blocks and Care Buses are functioning in practice, identify areas for improvement, and measure their contribution to achieving the program's objectives of recognition, redistribution, and reduction of caregiving responsibilities. This comprehensive evaluation is essential for refining the initiative, ensuring its sustainability, and demonstrating its value as a model for other cities and regions.

3.2.3 Implementation

3.2.3.1 Principal Parties and Partners

The Care Blocks initiative is built on the principle of shared responsibility, emphasizing the importance of collaboration in addressing the complex needs of both caregivers and care recipients. This principle is put into action through strong partnerships between public entities and private organizations, which combine their strengths and resources to establish a comprehensive care system. Private sector involvement has played a crucial role in expanding and improving the services provided by Care Blocks. For example, private companies have contributed resources to establish public and free washing facilities in each Care Block. These facilities are essential for relieving caregivers of daily burdens and providing them with access to necessary services, potentially saving up to one hour of handwashing each day. This contribution not only eases a significant aspect of caregivers' daily routines but also showcases how private sector participation directly enhances the quality of life for those involved in care work.

Furthermore, academia and philanthropic contributors are instrumental in the success and development of Care Blocks. Academic institutions have conducted studies and assessments that provide valuable insights into the effectiveness and impact of the initiative. Their research assists in refining the services offered, ensuring that they effectively meet the community's needs. On the other hand, donors provide financial support and resources, which are essential for sustaining and expanding the Care Blocks. Their contributions have facilitated the integration of advanced technologies and innovative solutions, further improving the functionality and efficiency of the Care Blocks. This multi-faceted approach to shared responsibility demonstrates how collaborative efforts can establish a more effective and sustainable care system. By leveraging the resources and expertise of different stakeholders, the Care Blocks initiative not only addresses immediate care needs but also cultivates a culture of collective support and innovation.

3.2.3.2 Resources for Implementation

Since the beginning of the Care Blocks initiative, a large portion of the total project cost, 98% to be exact, has been covered by significant public resources allocated through the City Development Plan 2020–2024. This demonstrates the city's dedication to meeting the care needs of its residents and establishing a supportive infrastructure for both caregivers and care recipients. The funding framework ensured that the main financial burden of the initiative was addressed through public channels, thereby enabling the effective establishment and operation of Care Blocks across Bogotá.

In addition to public funding, Care Blocks have received technical and financial assistance from various international cooperation agencies and academic institutions. For example, the 2021 Bloomberg Mayors Challenge Award recognized the initiative for its innovative approach to urban care systems, leading to international acclaim and further support and collaboration. Other significant supporters include the Carter Center's Transforming Lives Campaign, Open Society Foundations, and UN Women, all of whom have contributed their expertise and resources to strengthen the initiative.

Furthermore, international organizations like ECLAC, UNDP, UNHCR, IDB, and the World Bank have played crucial roles in providing additional funding and technical assistance. Their involvement has been essential in ensuring that Care

Blocks have the necessary tools and knowledge to operate effectively and sustainably. The private sector has also made valuable contributions through in-kind donations, providing essential goods and services to support the daily operations of Care Blocks. These contributions not only enhance the functionality of the Care Blocks but also demonstrate the private sector's commitment to social responsibility and community welfare. Overall, the combined effort involving public resources, international support, and private sector contributions has been pivotal in the successful implementation and ongoing development of Care Blocks, representing a significant advancement in urban care systems.

3.2.3.3 Tools and Methods Applied

The Bogotá Care Blocks initiative has achieved significant progress in improving the well-being of caregivers through the provision of two essential services. The public laundry service has effectively freed up 4,394 h for caregivers, equivalent to about 183 days previously spent handwashing clothes. This service not only reduces the physical burden on caregivers but also enhances their overall well-being by giving them valuable time for other activities. Similarly, the childcare service has released 7976 h, approximately 332 days, enabling caregivers to focus on personal development, income generation, or rest. Together, these services grant caregivers an extra four hours daily, a crucial amount of time that can be utilized for education, economic activities, or relaxation, significantly enhancing their quality of life.

The Care Blocks initiative closely aligns with several Sustainable Development Goals (SDGs), particularly emphasizing SDG #5 (gender equality) and its target 5.4, which seeks to acknowledge and value unpaid care work. Additionally, it supports SDG #3 (good health and well-being), SDG #11 (sustainable cities and communities), SDG #16 (peace, justice, and strong institutions), and SDG #17 (partnerships for the goals). By the target year of 2023, Bogotá aims to have 20 operational Care Blocks, with plans to increase this to 45 by 2035. This ambitious expansion is guided by Bogotá's Urban Master Plan, which provides a strategic framework for the development and implementation of additional Care Blocks. Furthermore, the recent approval of the Care System City Law by the City Council in March 2023 underscores a commitment to sustaining the program beyond current administrative terms, ensuring long-term stability and impact.

3.3 Obstacles and Solutions

The Care Blocks initiative in Bogotá marked a groundbreaking endeavor in urban policy, working at a level and intricacy unexplored in a city setting before. This ambitious program encountered several notable hurdles during its execution. One of the primary challenges was the impact of the COVID-19 pandemic, which worsened the already substantial burden of unpaid care work on caregivers. The pandemic heightened the need for support systems like Care Blocks, rendering the initiative even more crucial as caregivers grappled with heightened responsibilities and stress. Securing funding approval for such an innovative program also proved to be a significant obstacle. Obtaining the necessary financial support involved navigating the legislative process and persuading stakeholders of the program's enduring value, necessitating thorough planning and advocacy to ensure that the budget allocation met the program's extensive needs.

Another substantial challenge was coordinating among the 13 city government entities involved in the Care Blocks initiative. Traditionally, these entities operated independently, each managing their own services in isolation. Unifying their efforts into a collaborative approach necessitated overcoming institutional resistance and establishing new procedures and rules for joint operation. This transition called for extensive negotiation and reorganization to guarantee effective collaboration.

Moreover, creating a comprehensive data collection system to evaluate the impact of Care Blocks was critical yet difficult. The requirement for accurate and integrated data across various entities called for the development of a robust system capable of capturing and analyzing information to assess the program's effectiveness. Addressing these challenges was crucial for the successful implementation and sustainability of the Care Blocks initiative, underscoring its eventual success as a remarkable achievement in urban innovation and social policy.

3.3.1 Achieved Outcome

Since it was introduced in October 2020, the Care Blocks initiative has significantly impacted more than 297,000 caregivers and their families by providing over 1,900,000 services. This comprehensive support system has led to numerous positive changes within the community. For example, 849 women have successfully graduated from high school, showcasing the program's role in improving educational opportunities for caregivers. Moreover, over 12,000 individuals have received technical education, equipping them with valuable skills and qualifications that enhance employment prospects and career growth.

Additionally, the Care Blocks initiative has facilitated access to business and entrepreneurship education for over 28,000 participants. This educational support empowers caregivers to explore entrepreneurial ventures, promoting economic independence and contributing to local economic development. By offering these opportunities, the initiative not only enhances the personal and professional lives of individuals but also strengthens the broader community. The success of Care Blocks in providing such extensive services highlights its crucial role in addressing systemic issues of inequality and supporting caregivers in their personal and professional development. The program's accomplishments demonstrate its dedication to improving educational achievement, empowering economic independence, and ultimately transforming the lives of those who play a pivotal role in their families and communities.

3.4 Lessons Learned

The Bogotá City Government previously operated through separate departments, resulting in a fragmented approach to service delivery. Each department managed its own services and data, making it difficult to aggregate or analyze information. To address these challenges, Bogotá implemented a unified, interoperable data system, enabling a more coordinated approach to governance.

The COVID-19 pandemic emphasized the urgency of implementing Care Blocks, especially in supporting unpaid care work. The number of women engaged in unpaid care work increased from 800,000 to 1.2 million, highlighting the need for comprehensive support systems. Care Blocks thus became crucial in addressing immediate needs, promoting gender equality, and aiding economic recovery.

Care Blocks provide a replicable and adaptable model for urban and rural contexts, utilizing existing public infrastructure and an "anchor building" to bring together caregivers and care recipients. In rural and peripheral areas with limited service access, Care Buses were introduced to ensure that caregivers have access to essential services, addressing challenges such as limited access to basic utilities.

Related SDGs: Goals 3, 5, 11, 16, and 17.

Acknowledgements We explicitly thank the project owners based in Bogotá, Colombia for their participation and the provision of materials for the 6th Guangzhou International Award for Urban Innovation. We also congratulate them for their achievements in the 2023 cycle of awards.

The above images are provided to us by the project owners for their submission.

Chapter 4
Building Resilient Food Systems in Cape Town, South Africa

Abstract The food systems in Cape Town are complex and governed in a fragmented way, even though food is acknowledged as a constitutional right. The City has identified more than 40 interconnected actions related to food systems, covering areas such as informal trade, markets, environmental health, spatial planning, urban health, and particularly, resilience. Recent events like drought, COVID-19, load shedding, and minibus taxi strikes have highlighted the critical need for a comprehensive food systems approach that recognizes the growing frequency and impact of such disruptions on food access, especially for vulnerable communities. The City's approach takes a holistic view, aiming to integrate and leverage existing responsibilities through collaborative partnerships within the municipality and with external organizations. This strategy seeks to address food system challenges by promoting resilience, enhancing coordination across various sectors and stakeholders, and fostering a more cohesive response to future shocks and pressures while improving long-term sustainability and equity. The initiative is related to SDGs 2, 3, 11, and 12.

Keywords Cape town · Food systems · Resilience · Urban health · Spatial planning · Food access · Collaborative partnerships

4.1 Introduction to Cape Town, South Africa

Cape Town, South Africa's legislative capital and one of its most lively cities, symbolizes economic vitality, cultural variety, and natural splendor. With a population of about 4.8 million people, Cape Town stands as the second-largest city in South Africa and serves as a pivotal center for both the national and regional economies. The city's population has been steadily increasing at a rate of 1.85% per year, indicating its growing appeal as a destination for both residents and migrants in search of economic opportunities and a superior quality of life. Covering an expansive area of 2446 square kilometers, Cape Town has a population density of approximately 1638 people per square kilometer. This relatively modest density, in comparison to other

A. Cheshmehzangi et al., *People-Centered Innovation*, Urban Sustainability,
https://doi.org/10.1007/978-981-96-5230-3_4

major cities worldwide, underlines the city's distinctive urban layout, featuring extensive suburbs, green areas, and preserved natural spaces that significantly contribute to its overall desirability. Despite its vast expanse, the city's infrastructure and services are supported by a robust economic foundation, resulting in a high GDP per capita of $137,234, one of the highest in Africa. This economic prosperity is primarily driven by three key sectors: tourism, finance, and business services.

Tourism is a cornerstone of Cape Town's economy, with the city being a leading tourist destination on the African continent. Renowned for its stunning landscapes, including the iconic Table Mountain, Cape Town draws millions of visitors annually. The city's diverse cultural heritage and natural allure make it a magnet for both domestic and international tourists. Tourism not only brings in substantial revenue but also generates numerous employment opportunities, contributing to the livelihoods of many Capetonians. The finance and business services sectors also play a crucial role in Cape Town's economic success. The city is home to various financial institutions, such as banks, insurance companies, and investment firms, offering essential services not only to the local economy but also to the broader South African and African markets. Cape Town's status as a financial hub is bolstered by its well-established infrastructure, skilled workforce, and favorable business environment, attracting both domestic and international investments. Business services, including legal, accounting, and consulting firms, further enhance the city's economic standing by providing a range of professional services integral to the functioning of a modern economy.

In spite of its economic progress, Cape Town is confronted with significant difficulties, particularly concerning social inequality. The city possesses a GINI index of 0.62, indicating a high degree of income inequality. This evident gap mirrors the wider socio-economic disparities within South Africa, where wealth and resources are distributed unevenly among various racial and socio-economic groups. In Cape Town, this inequality is evident in multiple aspects, including differences in housing, education, healthcare, and access to essential services. Despite the existence of affluent neighborhoods with top-notch facilities, the city also contends with areas characterized by poverty, unemployment, and inadequate infrastructure. The spatial and economic inequalities in Cape Town are, to some extent, a legacy of the apartheid era, during which urban planning and development were tailored to segregate communities based on racial lines. These historical divisions continue to influence the city's urban layout, posing challenges for inclusive and fair development. Tackling these inequalities remains a key focus for the city's government, which is actively working to foster social harmony and economic inclusion through various policies and initiatives.

In recent years, Cape Town has also grappled with environmental challenges, most notably water scarcity. The city's dependence on a limited number of water sources, coupled with prolonged droughts, has made water management a critical concern. The water crisis of 2017–2018, when Cape Town came dangerously close to running out of water, emphasized the necessity for sustainable resource management and climate resilience. In response, the city has rolled out a variety of measures, including water conservation campaigns, infrastructure investments, and the diversification of water sources to ensure long-term water security.

The future development of Cape Town will hinge on its ability to harmonize economic growth with social justice and environmental sustainability. The city's leadership is increasingly concentrated on building a more inclusive economy that offers opportunities for all residents, while also safeguarding the natural environment that gives Cape Town its distinctiveness. Efforts such as the establishment of affordable housing, the expansion of public transportation, and the promotion of renewable energy are part of this broader strategy to construct a resilient and sustainable city.

As Cape Town continues to expand and transform, it retains its status as a city of contradictions—where prosperity and poverty coexist, and where natural beauty and urban challenges intersect. The city's capacity to navigate these complexities will shape its direction in the years to come. Nonetheless, with its robust economic foundation, lively cultural scene, and commitment to innovation and sustainability, Cape Town is well-placed to address these challenges and sustain its development as one of Africa's premier cities.

4.2 Background and Origins

The initiative concerning food systems in Cape Town is deeply ingrained in the city's broader dedication to establishing resilience, as detailed in the Cape Town Resilience Strategy put into effect in 2019. This strategy acknowledged the vital significance of food systems as a fundamental element of urban resilience, especially in a city such as Cape Town, where socio-economic inequalities and environmental obstacles are clearly evident. Post the approval of the Resilience Strategy, considerable endeavors were undertaken to formulate a comprehensive Food Systems Programme, which rapidly emerged as one of the flagship actions within the strategy.

The formulation of the Food Systems Programme was not just a reaction to local requirements but also in accordance with global commitments, notably the Milan Urban Food Policy Pact, of which Cape Town is a participant. This pact, endorsed by more than 200 cities globally, emphasizes the role of urban food systems in advancing sustainable development and addressing food security, nutrition, and climate resilience. By aligning the Food Systems Programme with this international framework, Cape Town has established itself as a frontrunner in urban food policy, actively contributing to the worldwide conversation on sustainable food systems. The alignment of the Food Systems Programme with the United Nations Sustainable Development Goals (SDGs) further highlights its significance. By integrating goals that contribute to SDG 2 (Zero Hunger), SDG 11 (Sustainable Cities and Communities), and SDG 13 (Climate Action), among others, the programme reflects a comprehensive approach to addressing food system challenges within the broader context of sustainable development.

The advancement of the program is evident in its formal inclusion in Cape Town's five-year Integrated Development Plan (IDP), a crucial document guiding the city's strategic planning and resource allocation. This integration signifies a noteworthy milestone, indicating that food systems are no longer perceived as a peripheral issue but as a central component of the city's development agenda. By integrating the Food Systems Programme within the IDP, Cape Town has guaranteed that it will receive the essential attention, resources, and inter-departmental cooperation required to tackle the intricate and interrelated challenges of urban food security and resilience.

Cape Town became involved in the Rockefeller 100 Resilient Cities Programme during a crucial period when the city was facing its most severe drought ever. This intense water scarcity emphasized the city's weaknesses and emphasized the urgent need for a more comprehensive approach to resilience. As part of participating in the program, Cape Town conducted a thorough assessment of the city's strengths and weaknesses across various sectors using the City Resilience Index benchmarking process. This process revealed food insecurity as one of the main vulnerabilities, which was worsened by the drought and posed a threat to the well-being of the city's most vulnerable populations.

In response to this, a food systems program was suggested as a strategic measure to address the underlying causes of food insecurity and its associated health impacts. The program acknowledges the prevalent dual burden of disease in Cape Town, where food insecurity not only leads to child wasting and stunting but also contributes to overweight and obesity, both of which drive the high incidence of non-communicable diseases (NCDs). This dual challenge necessitates a nuanced and multifaceted approach, which the city has embraced despite the absence of a clear local government mandate on food security. Traditionally, local governments have mainly participated in food-related initiatives through humanitarian relief efforts, such as distributing food parcels or promoting community food gardens. However, Cape Town's food systems program extends beyond these immediate relief efforts, adopting a more systemic approach to improving food access.

This systemic approach utilizes existing city mandates, including markets and trading, spatial planning, and health initiatives, to address food security more comprehensively. The program is centered on key themes such as resilience, governance, food environments, health, and the economy. Significant emphasis has been placed on governance and developing institutional knowledge within the city administration about the interlinked nature of food systems activities and their outcomes. By fostering a deeper understanding of how various city functions contribute to or detract from food security, Cape Town aims to build a more resilient and equitable food system that can withstand future shocks and stresses, thereby ensuring that all residents have access to nutritious and sufficient food.

4.2.1 Key Innovative Aspects

The testing phase in the beginning allowed the city to improve its strategy and show the importance of focusing on food systems. Ever since its official adoption in 2020, the program has been fully integrated into the City's long-term plan, particularly within the Integrated Development Plan (IDP) and the Metropolitan Spatial Development Framework (MSDF). This incorporation is a major success, as it guarantees that food systems considerations are now a fundamental part of the city's strategic planning for the future. Due to the extensive impact of food systems on various aspects of the city's operations, the program places great emphasis on advocacy and knowledge-sharing. To achieve this, Cape Town established a Food Systems Working Group, which is widely acknowledged as a best practice model within the city. This group brings together both internal and external stakeholders to exchange knowledge, establish connections, and align the city's activities with food-related goals. The working group plays a vital role in promoting collaboration and ensuring that food systems are considered in a comprehensive and integrated manner across different departments and sectors.

In addition to the working group, the city hosts various initiatives to encourage food systems research and coordination. These efforts include the Food Indaba, a city-wide meeting that unites key players to discuss and develop strategies for food-related issues, as well as the Food Systems Research Community of Practice, which concentrates on coordinating and advancing research endeavors. The city also takes a central role in organizing events for food systems actors, which work to align resources and efforts across various sectors, ensuring that the combined impact of these activities is optimized.

4.2.2 Desired Outcome and Measurement

4.2.2.1 Desired Outcome

The City's Food Systems Programme aims to achieve a transformative and long-lasting impact by embedding 'food' as a central component of City policy and strategy. This integration seeks to enhance governance and ensure effective implementation across various municipal departments and initiatives. The ongoing work is supported by a dedicated Implementation Programme that is set to continue for another 12 months, focusing on immediate actions and building momentum for broader systemic change. However, the programme's vision extends beyond this timeframe, aligning with the City's 5-year Integrated Development Plan, which runs until 2027. The desired outcome is to initiate a systems change that will have a sustainable impact on the city's food security, resilience, and overall well-being. By fostering cross-departmental collaboration, enhancing governance structures, and continuously building knowledge and capacity within the organization, the City envisions a future where food systems are robust, resilient, and capable of withstanding shocks and stresses. This long-term approach is designed to ensure that the City's food systems are not only responsive to current challenges but are also adaptable to future demands, ultimately contributing to a more equitable and sustainable urban environment.

4.2.2.2 Measurement

The annual Food Report serves as a crucial tool for monitoring and evaluating the City's activities within the Food System (Fig. 4.1). Given the comprehensive, system-based approach, the report aggregates data and activities from multiple departments, offering a high-level overview of the progress and challenges faced in implementing the Food Systems Programme. This broad scope ensures that the report captures the multifaceted nature of food systems governance, though it necessarily presents a summary rather than granular detail. Nevertheless, the report includes key metrics such as health data, reflecting the impact of food insecurity and dietary patterns on public health. Additionally, the report tracks food trade data, providing insights into the availability, distribution, and economic impact of food within the city. Over time, the Food Report aims to refine its indicators and data collection processes, incorporating more detailed and real-time data to offer a more nuanced understanding of the city's food system, its vulnerabilities, and its progress toward resilience and sustainability. This continuous improvement in measurement is essential for aligning the City's efforts with its long-term goals and ensuring that the Food Systems Programme remains responsive to evolving challenges.

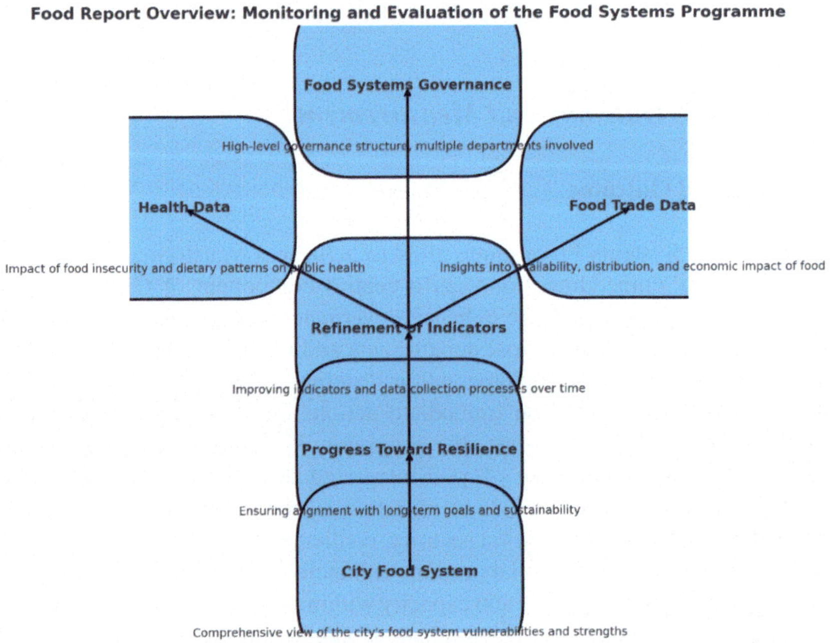

Fig. 4.1 Summary of measurement for the Food Report overview. It highlights the different components such as food systems governance, health data, food trade data, and the continuous refinement of indicators. The arrows indicate the flow of activities and their connections within the larger food systems framework. The image is developed by the Authors using AI-assisted tools and materials provided by the project owners

4.2.3 *Implementation*

4.2.3.1 Principal Parties and Partners

The City of Cape Town drives a collaborative network involving provincial government levels, universities, NGOs, and researchers as part of the initiative. Each partner has a vital role in improving food system outcomes through various projects and research support within this extensive network. The initiative stands out for its adaptable and non-hierarchical approach in maintaining partnerships and processes, allowing all stakeholders to contribute significantly without being constrained by rigid structures.

The small team's success lies in strategic networking both within the city and with external partners. This networking enables the efficient flow of knowledge and resources to address food system challenges. The team's close relationships with researchers and other organizations are particularly valuable, providing quick access to relevant data, especially during crises. This timely access to data ensures informed responses to emerging challenges, enhancing the city's resilience and adaptability.

The initiative's partnerships go beyond immediate project-based collaborations, contributing to long-term improvements in food governance, research, and policy development. Close collaboration with provincial governments ensures that food system strategies align with broader regional goals, creating a more cohesive approach to food security. Universities and researchers provide critical data and analytical support, while NGOs offer grassroots perspectives and direct community engagement, ensuring that the voices of those most affected by food insecurity are heard and considered in decision-making. In summary, the initiative's partnership model reflects a collaborative, innovative approach to governance that prioritizes flexibility, inclusivity, and responsiveness, setting a standard for effectively managing urban food systems in the face of ongoing challenges.

4.2.3.2 Resources for Implementation

Cape Town's Food Systems Programme relies on a combination of municipal funding and external grants to improve food systems despite limited resources. The city's financial support primarily covers salaries for the programme's core team, ensuring consistent operation. In addition, the initiative benefits from an EU Horizon grant provided through ICLEI's AfriFOODLinks programme, supporting market-focused projects until 2025. This external funding enables the city to explore and develop innovative solutions in food market operations, crucial to the programme's overall success.

Cape Town not only makes financial contributions but also emphasizes building knowledge and capacity within the food systems space. This commitment is evident in its practice of hosting interns who contribute to the programme while gaining valuable experience. By fostering an environment of learning and knowledge exchange, the city enhances its internal capabilities and helps cultivate the next generation of professionals in urban food systems management. Both financial and intellectual resources are critical to the ongoing success of Cape Town's Food Systems Programme. The combination of direct city funding, strategic external grants, and a focus on knowledge-building allows the initiative to maintain its momentum and continue making strides in improving food outcomes in the city. Through careful resource management, Cape Town navigates the complexities of food systems governance and ensures that its efforts are sustainable, impactful, and aligned with broader goals of urban resilience and social equity.

4.2.3.3 Tools and Methods Applied

The Food Systems Programme in Cape Town takes a comprehensive approach, using various methods to engage stakeholders, build networks, and influence policy. By integrating the programme into the City's Integrated Development Plan (IDP) for a five-year period, Cape Town shows commitment to prioritizing food systems as part of its broader urban development strategy, emphasizing sustainable food access for all residents. An important achievement for the programme is securing resources through the AfriFOODLinks initiative, funded by an EU Horizon grant via ICLEI. This partnership is crucial for addressing informal food access, which is how a majority of Capetonians obtain their food. By focusing on markets and trade, the programme aims to improve the availability and quality of food in informal sectors, thus enhancing food security for a substantial portion of the population.

Significant progress has been made in food waste management. By recognizing the potential impact of solar-powered cold storage, the city is exploring this technology to extend the shelf life of fresh produce, reducing food waste and improving food quality. This approach not only addresses waste reduction but also supports the broader goals of sustainability and resilience within the food system. The city is also exploring the development of nutrition hubs, including early childhood development centers, community kitchens, and food gardens. These hubs are in the early stages but are gaining momentum, offering a holistic approach to addressing nutritional needs at the community level by providing spaces for education, access to healthy food, and community engagement.

Regular participation in Food Dialogues has further strengthened the City's position as a leader in food systems thinking. These dialogues provide a platform for building networks, sharing knowledge, and influencing food-related policies and practices. By engaging with a wide range of stakeholders, including children and youth, the City is fostering a culture of inclusion and education vital for the long-term sustainability of its food systems. Through these efforts, Cape Town is addressing immediate food access and security challenges while laying the groundwork for a more resilient and equitable food system. The strategic partnerships, innovative projects, and commitment to policy integration ensure that the City's Food Systems Programme is well-positioned to meet the needs of its diverse population and contribute to broader urban resilience goals.

4.3 Obstacles and Solutions

The City of Cape Town's Food Systems Programme faces a significant obstacle due to the narrow view of food systems, often perceived as limited to food gardens. While food gardens contribute to local food production and community engagement, they are just a small part of the wider food system. Changing this perception has been quite challenging, especially at a high political level. This limited perspective overlooks the broader responsibilities of local government, such as food access, distribution, waste management, and integrating food into urban planning and policy frameworks, which are all systemic issues.

The narrow perspective hampers the programme's ability to comprehensively address food security and resilience. Furthermore, the Food Systems team is relatively small, and their limited resources directly impact their productivity. The team's capacity to expand initiatives, build networks, and influence policy is restricted by these resource constraints. As a result, the programme struggles to scale its efforts and fully integrate food systems thinking into the broader urban governance landscape. Overcoming these challenges will require not only increasing the team's resources but also fostering a more nuanced understanding of food systems at the political level, ensuring that local government approaches food security in a holistic and systemic manner.

4.3.1 Achieved Outcome

The City of Cape Town has recently shown a significant shift towards acknowledging the interconnection of food systems with broader urban policies and strategies. This evolving perspective is evident through various key initiatives and events that emphasize the integration of food systems into different aspects of city governance. For example, a notable instance is the inclusion of a food systems breakaway session at this year's Informal Trade Summit, which was arranged at the invitation of economic development colleagues. This session highlighted the increasing recognition of food systems as essential to informal trade and economic development.

Furthermore, the incorporation of food systems in the City's Integrated Development Plan and the Metropolitan Spatial Development Framework demonstrates a formal commitment to integrating food-related considerations into long-term urban planning. These planning documents offer a structured approach to incorporating food systems into broader city strategies, ensuring that food security and resilience are systematically addressed.

Looking forward, the City is also getting ready for a Water and Food event scheduled for November this year, aiming to further explore and advance the connection between water management and food systems. This event presents an opportunity to deepen the discussion on how these critical areas intersect and to develop collaborative solutions that enhance urban resilience. Collectively, these efforts demonstrate a growing acknowledgment of the importance of food systems in shaping sustainable and resilient urban environments, representing a significant advancement in the integration of food considerations into city-wide planning and policy.

4.4 Lessons Learned

The COVID-19 pandemic posed considerable difficulties in delivering humanitarian aid to families and determining the most effective ways to help those in need on the ground. The experience highlighted a number of important lessons, especially regarding the role of local partners and the impact of different types of assistance. Local partners, deeply rooted in affected communities, proved to be extremely valuable during the crisis. Their in-depth understanding of the communities they serve allowed them to identify those in need more accurately and navigate complex urban and societal systems more efficiently than larger, less flexible organizations. Keeping updated databases of these local partners has become a vital tool for quickly and efficiently mobilizing relief efforts during emergencies.

Another significant insight was the preference for food vouchers over food parcels. Vouchers provided a more adaptable and user-friendly solution compared to the logistical challenges associated with distributing food parcels. The vouchers could be used within local economies, supporting local businesses and maintaining economic activity within communities. This method helped avoid potential destabilization of local systems caused by external aid and contributed to the resilience of local economies. The collaboration with research advisors played a critical role in shaping interventions that were considerate of local dynamics. This partnership ensured that decisions were well-informed and nuanced, preventing interventions from unintentionally disrupting existing community structures.

Building on these experiences, the initiative now focuses on integrating social infrastructure into the broader structure of the City. This approach aims to enhance the City's ability to respond to future shocks by integrating lessons learned from the pandemic. Key lessons for other cities include the importance of involving specialized advisors, the necessity of robust data and communication tools for planning and response, and the potential benefits of embracing non-traditional policy approaches. By systematizing these elements, the City aims to improve its capacity to address future crises more effectively and sustain community resilience.

Related SDGs: Goals 2, 3, 11, and 12.

Acknowledgements We explicitly thank the project owners based in Cape Town, South Africa for their participation and the provision of materials for the 6th Guangzhou International Award for Urban Innovation. We also congratulate them for their achievements in the 2023 cycle of awards.

Above is an image provided to us by the project owners for their submission.

Chapter 5
Ancient Engineering of Sustainable Water Management Meets Urban Innovation Through Citizens' Participation in Chalandri, Greece

Abstract The Cultural H.ID.RA.N.T. project (Cultural Hidden IDentities ReAppear through Networks of Water, henceforth CH) is dedicated to activating local Hidden Heritage Potential and resources—both tangible and intangible, cultural and natural—to enhance Local Wellbeing. This innovative initiative has both local and inter-local outreach, using water as a medium to connect cultural heritage with urban revitalization. By reintroducing the Hadrian Aqueduct, an ancient Roman infrastructure, CH transcends the conventional "sightseeing" approach. It integrates this historic monument as a contemporary landmark, local memory, and underutilized water resource within the urban fabric of Chalandri. The project creates a paradigm shift by enhancing the cultural branding of Chalandri in contrast to the heritage focus of central Athens. CH promotes green urban policies and sustainable water management by regenerating four neighborhoods in Chalandri, transforming them into quality green public spaces. The Hadrian Aqueduct is repurposed for irrigation, replacing expensive and energy-consuming potable water. Through active community engagement, CH revitalizes Chalandri's communities, strengthens the local economy, and fosters endogenous and resilient development. The initiative is related to SDGs 6, 9, 11, 12, 13, and 17.

Keywords Cultural heritage · Urban revitalization · Sustainable water management · Green urban policies · Community engagement · Heritage conservation · Local wellbeing

5.1 Introduction to Chalandri, Greece

Chalandri, an energetic municipality located in the northern region of Athens, Greece, is a city that combines historical significance with contemporary vitality. With a population of 77,108, Chalandri is a thriving urban area characterized by a notable population growth rate of 3.80%. Encompassing a significant land area of 10,805 square kilometers, Chalandri is renowned for its densely populated and lively environment, with a population density of 7,140 people per square kilometer. The city's

A. Cheshmehzangi et al., *People-Centered Innovation*, Urban Sustainability,
https://doi.org/10.1007/978-981-96-5230-3_5

economic prosperity is primarily driven by its strong service and leisure industries, which play a crucial role in its economic landscape.

Chalandri's strategic position within the Greater Athens area contributes significantly to its economic and social vibrancy. As an integral part of the Athens metropolitan area, Chalandri benefits from its proximity to the Greek capital, while also fostering a unique local identity that distinguishes it from its neighboring districts. The city's economy is bolstered by a diverse range of service-based businesses, including retail, hospitality, and professional services, alongside a growing leisure industry that caters to both residents and visitors. Historically, Chalandri has evolved from a small village to a bustling urban center, mirroring the broader transformation of the Athens metropolitan region. This evolution has been characterized by rapid urbanization and modernization, which have shaped the city's contemporary character. Despite these changes, Chalandri has managed to preserve elements of its historical and cultural heritage, contributing to its unique allure and appeal.

One of the most significant aspects of Chalandri's historical and cultural identity is its link to ancient Greece. The city is home to several archaeological sites and historical landmarks that reflect its rich past. Among these is the Hadrian Aqueduct, a remarkable relic from the Roman era that has played a crucial role in shaping the city's historical landscape. This subterranean infrastructure is a testament to the ingenuity of ancient engineering and remains an important symbol of Chalandri's historical heritage.

In recent years, Chalandri has adopted a forward-thinking approach to urban development, focusing on sustainability and quality of life. The city's dedication to enhancing its public spaces and green areas has been a central theme in its urban planning efforts. This is evident in the development of new parks, recreational facilities, and pedestrian-friendly areas that aim to improve the overall well-being of its residents. The economic growth of the city has been driven by its thriving service and leisure sectors, offering a wide range of amenities to meet the needs of its diverse population. Chalandri boasts vibrant entertainment venues and bustling shopping centers, making it an attractive urban environment for both locals and visitors. The city's strong emphasis on leisure and recreation has nurtured a sense of community, with numerous cultural events and festivals celebrating its rich heritage and contemporary culture. Chalandri's growth and progress are closely tied to its strategic planning and investment in infrastructure. The city has embarked on various initiatives to improve its connectivity and accessibility, including enhancements to transportation networks and public services. These efforts are aimed at creating a more integrated and efficient urban environment to support the needs of its growing population.

Despite its advancements, Chalandri faces challenges typical of rapidly urbanizing cities, such as traffic congestion, environmental sustainability, and the preservation of historical sites. Addressing these challenges involves innovative solutions and community engagement to ensure sustainable and inclusive development. Looking ahead, Chalandri aims to maintain its vibrant character while meeting the evolving needs of its residents. The city is committed to enhancing the quality of life by investing in public spaces, cultural initiatives, and sustainable practices. By leveraging its historical heritage and embracing modern advancements, Chalandri strives to create a balanced and thriving urban environment that reflects its unique identity.

In summary, Chalandri is a dynamic and evolving city that seamlessly blends its rich historical heritage with contemporary urban living. With a strong foundation in its service and leisure industries, the city is poised to continue its growth while preserving its distinctive character. Through strategic planning and community engagement, Chalandri is working towards a future that embraces both its historical legacy and modern aspirations, ensuring a prosperous and vibrant environment for future generations.

5.2 Background and Origins

Chalandri has undergone significant changes in recent years due to a growing emphasis on service and leisure industries, leading to both economic prosperity and strain on the city's urban, cultural, and natural resources. The increase in population has created challenges such as traffic congestion and pressure on infrastructure, especially during weekends when around 7,000 visitors influx into the city, disrupting the local community and access to public spaces. The city's infrastructure and public areas have been strained by the growing demand, impacting the overall quality of life and diminishing the local identity and sense of unity. To tackle these issues, the CH project, part of the U.I.A. program, which is supported by the ERDF, has been launched to revitalize Chalandri's cultural and natural assets, with a focus on reintegrating the Hadrian Aqueduct, using the city's historical infrastructure to mitigate the negative effects of rapid development and tourism while enhancing cultural heritage and community integration.

5.2.1 Key Innovative Aspects

Chalandri's commitment to urban development is characterized by an integrated approach that considers cultural, social, environmental, and economic aspects. This comprehensive strategy aims to balance various facets of city life, reflecting a broader view of urban resilience and sustainability. The core of this approach lies in acknowledging that addressing urban challenges requires a multifaceted strategy that considers the needs and contributions of different sectors. A significant aspect of Chalandri's approach is its focus on participatory planning and governance, giving communities the power to take charge of public resources, fostering a sense of responsibility and involvement among residents. By involving local stakeholders in decision-making processes, the city ensures that development projects resonate with the needs and ambitions of its people. This participatory approach not only enhances the effectiveness of urban interventions but also builds stronger community bonds and trust.

Another fundamental element of Chalandri's innovation is the revitalization of cultural heritage. Unlike traditional preservation approaches that primarily focus on tourism and conservation, Chalandri's strategy aims to reintegrate cultural heritage into the daily life of the city. For example, the reintroduction of the Hadrian Aqueduct serves not only as a historical attraction but also as a functional component of the city's infrastructure. This reimagining of heritage creates a dynamic interplay between historical significance and contemporary utility, breathing new life into dormant resources and increasing their relevance in the modern context. Sustainable and circular economies are also central to Chalandri's strategy. The city is committed to reusing idle resources as shared assets, promoting resilience and internal development. By embracing a circular economy model, Chalandri aims to minimize waste and maximize the value obtained from existing resources. This approach not only supports environmental sustainability but also fosters local economic growth and stability. The integration of sustainable practices into urban development showcases a dedication to long-term ecological and economic well-being.

The governing structure that supports Chalandri's initiatives is both well-organized and adaptable. A steering committee plays a vital role in making significant decisions and monitoring progress on a quarterly basis, ensuring that the project remains aligned with its objectives and addresses any emerging issues promptly. Additionally, weekly operational meetings facilitate ongoing coordination and troubleshooting among team members, which are essential for maintaining momentum and ensuring effective management of all aspects of the project. The city promotes collaboration by organizing working meetings and workshops, which act as platforms for consultation and joint planning of activities. These sessions are crucial for involving different stakeholders, such as community members, experts, and partners, in gathering input and coordinating efforts. Through nurturing a collaborative environment, Chalandri improves the quality and impact of its initiatives.

The project's dedication is formalized through a range of agreements and declarations. Partners have developed a declaration of intent to demonstrate their commitment to the project's success. Furthermore, cooperation memorandums and charters between citizen groups, the Cultural H.ID.RA.N.T. project, and the Chalandri Municipality outline the roles and responsibilities of each involved party in jointly managing the project's outcomes. These formal agreements establish a framework for collaboration and ensure that all stakeholders are aligned in their endeavors to achieve the project's objectives.

In essence, Chalandri's pioneering approach to urban development is defined by its integration across sectors, participatory governance, innovative heritage renewal, and dedication to sustainable and circular economies. The city's governance structure, which includes the steering committee, operational meetings, and collaborative workshops, facilitates effective implementation and coordination. The formal agreements and declarations of commitment further strengthen the collaborative ethos propelling the project's success. Through these innovative strategies, Chalandri is leading the way for a more resilient, inclusive, and sustainable urban future.

5.2.2 *Desired Outcome and Measurement*

5.2.2.1 Desired Outcome

The CH project aims to bring about significant improvements to Chalandri's urban environment and community well-being by achieving various key outcomes. Enhancing the quality and accessibility of public green spaces is a primary objective, with the intent of creating a stronger sense of community and belonging among residents through their revitalization. Additionally, the project aims to showcase the city's rich cultural heritage by leveraging historical assets like the Hadrian Aqueduct to establish meaningful connections between past and present.

Another crucial goal is to establish a sustainable water infrastructure managed by citizens. This initiative is based on principles of circular and solidarity economies, focusing on resource efficiency and community-driven solutions. By implementing a participatory approach to urban resource management, the CH project seeks to empower residents to actively participate in the care of their environment. The project's timeline spans four years, concluded on June 30, 2024. Nevertheless, several aspects are already in operation. The participatory urban planning process, the HIDRANT festival, and the digital local history platform have been successfully completed and are currently in use. The Chalandri Adrian Community Network was established by December 2023, providing a structured forum for community engagement. Regeneration works and the non-potable water network were finalized in April 2024, at which point the new water delivery system commenced, further advancing the project's sustainability goals.

5.2.2.2 Measurement

The project "Cultural H.ID.RA.N.T." (CH) is supported by a thorough monitoring and evaluation (M&E) framework managed by Panteion University. This framework utilizes a combination of methods to evaluate the multifaceted impact of the initiative and assess "change" across different aspects. An essential part of this assessment is a two-phase survey on life satisfaction involving around 100 participants. The survey aims to establish the extent to which the project is linked to increased local well-being, with a goal of 50% of participants reporting high levels of life satisfaction associated with CH (Fig. 5.1).

In addition to overall life satisfaction, the project uses several specific indicators to assess its impact. These include measuring the residents' sense of connection to their heritage to determine how effectively the project promotes a feeling of cultural continuity and pride. The project also evaluates changes in the residents' sense of ownership and belonging to assess how the initiative impacts their emotional investment in public spaces and cultural assets.

Another important area of focus is skills development, particularly related to heritage co-governance and the organization of public events. This involves assessing

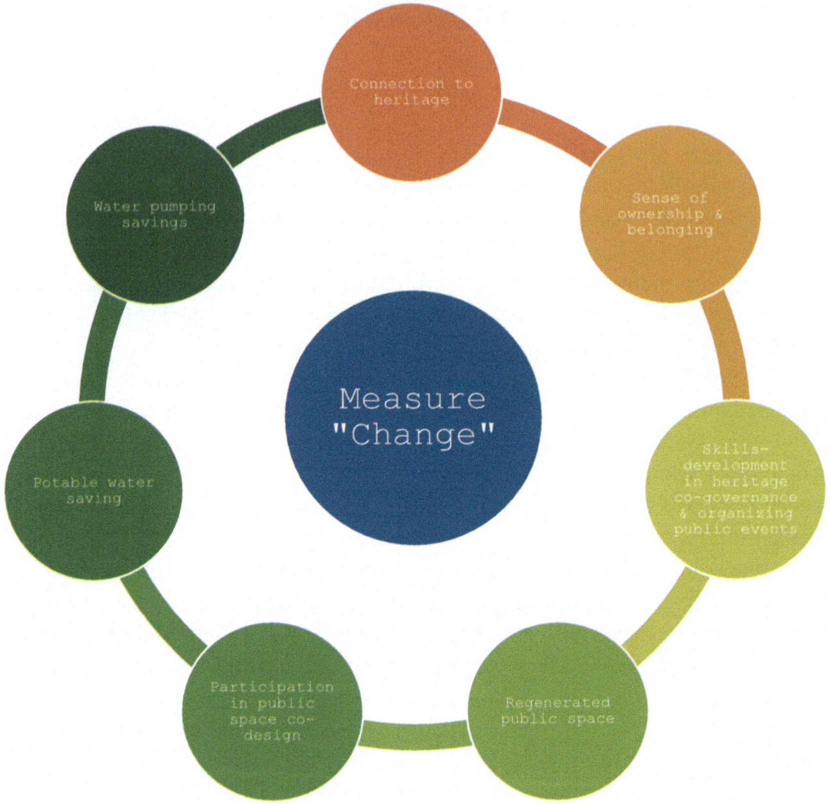

Fig. 5.1 Findings of measuring "Change" through the survey of 100 participants

how well the project improves local capacity to manage and celebrate heritage through skill-building workshops and participatory activities. The project measures the regeneration of public spaces to ensure that the physical enhancements align with the project's objectives, including the quality and usability of these areas for residents.

The project monitors participation in the co-design of public spaces to determine how effectively it engages community members in shaping their environment. This level of involvement is crucial in ensuring that public spaces meet the needs and preferences of the people who use them. The project also tracks water-related indicators, such as savings in potable water usage and reductions in water pumping costs, to assess the environmental impact of the new infrastructure and sustainable practices implemented through CH. In summary, the M&E processes are designed to provide a comprehensive understanding of how the CH project contributes to local well-being, cultural heritage, and environmental sustainability.

5.2.3 *Implementation*

5.2.3.1 Principal Parties and Partners

The CH project relies on a strong partnership model involving the public, private, and community sectors to achieve its objectives. Chalandri Municipality acts as the project manager, overseeing and coordinating activities to ensure compliance with municipal priorities and regulations. Each partner in the consortium leads specific work packages and tasks, using their expertise to advance different parts of the project. This approach ensures that organizations best suited to handle specific tasks are responsible for them, improving efficiency and effectiveness.

Stakeholders actively participate in both the design and implementation phases of the project. Their involvement is essential in shaping and carrying out the initiative to ensure that the results meet community needs and desires. By engaging in the project's development and operational phases, stakeholders also take responsibility for specific outcomes, fostering a sense of ownership and dedication to the project's success. This collaborative model not only optimizes resource allocation but also enhances the project's sustainability by establishing a strong foundation of shared goals and mutual commitment among all involved parties.

5.2.3.2 Resources for Implementation

The CH project receives funding from a variety of sources, including the ERDF via the UIA program, the Municipality of Chalandri, and the Greek Green Fund. These financial resources are essential for supporting the initiative's different aspects and implementing innovative solutions for urban regeneration and heritage revitalization.

In addition to financial support, the active involvement of local communities and stakeholders has been one of the project's most valuable strengths. The Oral History Group and the Rematia Stream SOS Association are crucial in co-designing, implementing, and maintaining the project's outcomes, ensuring its deep connection to the local context and genuine community needs. Engaging local residents and stakeholders has been vital in identifying and addressing these needs. By involving them, the project effectively utilizes human and cultural resources on-site. This collaborative approach not only increases the project's relevance and impact but also encourages ownership and commitment among participants. Through their contributions, these local actors ensure that the project's interventions are both meaningful and sustainable, maximizing the benefits for Chalandri's urban and cultural landscape.

5.2.3.3 Tools and Methods Applied

The CH initiative closely aligns with the main regional and RIS3 priorities, particularly focusing on Creative Blue and Sustainable Economy. An outstanding accomplishment is the advancement of cultural heritage awareness through various creative initiatives. With 350 items now housed in the Citizen-led Digital Local Archive, and the successful hosting of the HIDRANT Festivals three times, along with the launch of the Drop-A-Message digital game-app, all contributing to increased public engagement with Chalandri's cultural assets.

In the realm of natural heritage and sustainable water use, CH has surpassed its initial goals significantly. The number of water users has more than doubled from the initial target of 70 to 180, and Chalandri's green spaces have expanded by 300%, reflecting the project's substantial impact on urban environmental quality. The emphasis on participatory planning has been impactful too, with 1,500 square meters (7%) of regeneration achieved through community involvement. This process engaged 13 school communities and approximately 700 participants, ensuring that local needs and aspirations are taken into account.

The initiative has also promoted community building through the establishment of the Hadrian Community of Chalandri, an association dedicated to ongoing collaboration and engagement. The Hadrian Network digital platform further supports this endeavor by connecting individuals and organizations involved in the project, enabling communication and enhancing the overall impact of the initiative. These achievements demonstrate CH's success in integrating cultural and environmental priorities, promoting community engagement, and fostering sustainable development in Chalandri.

5.3 Obstacles and Solutions

The CH project faced significant hurdles, mainly related to bureaucracy. Securing approvals for technical design studies and navigating complex public procurement procedures posed major challenges. These issues caused delays in the project timeline, leading the leading partner to request and successfully obtain a one-year extension from the Urban Innovative Actions (UIA) program. This extension was crucial to accommodate the unforeseen delays and ensure the successful continuation of the project.

Another point of contention was the establishment of a distinct tariff system for non-potable water services, with implications for water resource management and public perception. There was also debate about integrating citizen participation in water resource management as a common good. To address these challenges, the City engaged in comprehensive discussions and consultations with EYDAP, the water operator and project partner, as well as specialists and urban authority representatives experienced in water co-management schemes. Through participatory workshops and collaborative dialogues, the City developed a two-level operational map, outlining how citizens can be involved in water resource stewardship at both local and trans-local levels. This provides a framework for effective and inclusive water management, aiming to balance regulatory requirements with community engagement and foster a more sustainable and participatory model for water resource management.

5.3.1 Achieved Outcome

The urban landscape of Chalandri has been greatly improved by the CH project, which combines cultural heritage with sustainable resource management. A significant accomplishment has been the restoration of the Hadrian Aqueduct as a heritage site and a practical water source. This restoration involved laying a 4,000-m pipe network, installing 100 smart meters, and deploying two municipal water tracks, directly benefiting around 170 residents. As a result, the aqueduct not only serves as a historic monument but also contributes to effective water management solutions.

The project has also encouraged community involvement and rejuvenation. By engaging approximately 1,000 residents in participatory activities, CH has nurtured a sense of ownership and strengthened community bonds. Moreover, the establishment of the first local history archive and digital platform, showcasing 350 original items, has provided a valuable resource for preserving and sharing Chalandri's cultural heritage. In terms of environmental impact, CH has made significant progress in promoting environmentally friendly urban policies. The initiative has achieved substantial water savings, amounting to 25,000 cubic meters annually, with 15,000 cubic meters of this being drinkable water. Additionally, the project has successfully regenerated 21,400 square meters of green space across four neighborhoods. These results emphasize CH's dedication to improving the quality of life for residents and the sustainability of urban resources, solidifying its position as a model for integrating cultural and environmental priorities in city planning.

5.4 Lessons Learned

The CH initiative has been strongly dedicated to tackling the climate emergency and improving public spaces in Chalandri. At the core of its approach is the revitalization of water infrastructure and the introduction of green spaces in urban areas, which together help create microclimates to combat the local impacts of climate change while enhancing the well-being of residents. By renovating outdated water systems and expanding green areas, the project aims to offset negative environmental effects and promote healthier urban environments.

The endeavor not only improves urban infrastructure but also encourages a culture of responsible resource management among residents. By involving communities in the maintenance of their technical, urban, and natural surroundings, CH promotes the adoption of sustainable behaviors and a deeper connection to local heritage.

This change in culture is crucial in nurturing a resilient and environmentally aware community. CH has taken inspiration from international models of participatory water management, such as those in Torino, Paris, and Terasa in Catalunya. The Municipality of Chalandri's participation in an inter-municipal consortium to expand these practices demonstrates its dedication to broader collaborative initiatives. The project has also gained recognition at a European level, taking part in Creative FLIP and presenting at six conferences, including one organized by GWOPA of UN Habitat. Additionally, CH is involved in urban networks like the HUB-IN Alliance and Cities FORUM, further cementing its role in shaping water and cultural policies.

As the project nears its completion, it is preparing for a series of end-of-project conferences and inter-local meetings with municipal authorities and stakeholders. These events will showcase the project's accomplishments, lessons learned, and innovations. A promotional handbook will be published to share best practices, addressed challenges, and developed solutions, ensuring that the knowledge gained from CH contributes to the broader conversation on sustainable urban development and climate resilience.

Related SDGs: Goals 6, 9, 11, 12, 13, and 17.

Acknowledgements We explicitly thank the project owners based in Chalandri, Greece for their participation and the provision of materials for the 6th Guangzhou International Award for Urban Innovation. We also congratulate them for their achievements in the 2023 cycle of awards.

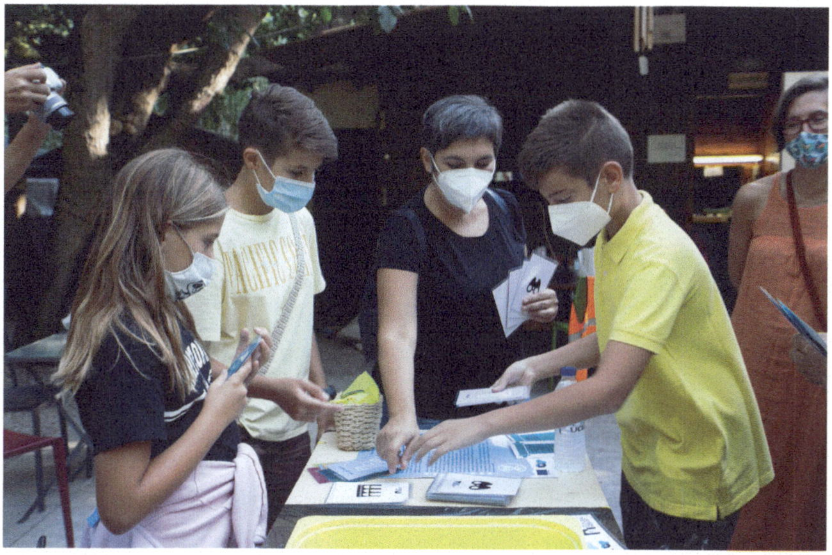

The above images are provided to us by the project owners for their submission.

Chapter 6
The Last Bastion for Protecting Human Dignity 'Gwangjuness Community Care' in Gwangju, Republic of Korea

Abstract In situations where individuals are unable to perform daily tasks due to health conditions, the assistance of others can become a crucial means of preserving their dignity. However, the current caregiving system often restricts support to those who meet specific criteria, such as age or income levels, and requires eligible individuals to actively apply for services. This approach can leave many in need without the necessary care. To address these limitations, the innovative "Gwangjuness Community Care" model has been developed. Unlike traditional systems, this model is designed to provide care to anyone in need, regardless of their age, income, or other qualifying factors. The initiative empowers care managers based in local dong-offices to proactively visit individuals in their homes, assess their situations, and apply for the appropriate services on their behalf. The Gwangjuness Community Care model has introduced 16 new care services, supported by an annual budget of $7.74 million. This financial commitment is aimed at ensuring that citizens facing economic challenges can access the care they require. To further enhance accessibility, a "Care Call" system has been launched, providing a single, streamlined application channel that simplifies the process for those seeking assistance. This comprehensive approach not only broadens the scope of care to include all who need it but also reduces barriers to accessing essential services, thereby enhancing the well-being and dignity of the city's most vulnerable residents. The initiative is related to SDGs 1, 2, 3, 5, 6, 7, 8, 10, 11, and 17.

Keywords Gwangjuness community care · Innovative care model · Caregiving system · Health and dignity · Social welfare · Proactive care management · Accessible care services

6.1 Introduction to Gwangju, Republic of Korea

Situated in the southwest region of the Korean Peninsula, Gwangju is a city that effectively combines contemporary innovation with a rich historical legacy. Gwangju, one of South Korea's major urban centers, has a population of 1,425,739 and a

A. Cheshmehzangi et al., *People-Centered Innovation*, Urban Sustainability,
https://doi.org/10.1007/978-981-96-5230-3_6

density of 2,944 people per square kilometer. It is well-known for both its economic accomplishments and its crucial role in the country's democratization movement.

Gwangju is particularly significant in Korean history because it was the center of the 1980 May 18 Gwangju Democratization Movement. This movement has been recognized by UNESCO's Memory of the World program as a pivotal event in the country's democratic struggle. The movement's spirit of resistance and the struggle for human rights have shaped the city's identity to this day, giving its citizens a strong sense of community and civic pride.

Gwangju's GDP per capita is high at $82,408, indicating that the city has maintained a strong economy despite a slight decline in population (growth rate of −0.57%). The city's strategic emphasis on innovation and technology is largely responsible for its economic resilience. Gwangju has developed into a center for innovative industries over time, especially in the areas of artificial intelligence (AI), renewable energy, and smart technology. The city's investment in R&D, which draws talent and companies from around the world, further demonstrates its dedication to promoting innovation. The development strategy of the city places a strong emphasis on sustainable growth and includes programs to support green technologies and lower carbon emissions. As a result, Gwangju is now recognized as a pioneer in the Korean peninsula's efforts to tackle climate change, setting an example for other cities in the vicinity.

The 501.1 square kilometer surface area of Gwangju serves as a blank canvas for urban planners who have skillfully combined contemporary infrastructure with the city's natural surroundings. The urban landscape is distinguished by a blend of modern architectural design and conserved historical landmarks, offering a superior standard of living to the inhabitants. The comparatively low GINI index of 0.33 for Gwangju suggests that wealth is distributed fairly evenly, which supports social stability and a strong sense of community. The city's numerous universities and cultural facilities demonstrate its dedication to education and culture. Numerous academic institutions, research centers, and cultural hubs can be found in Gwangju, including the well-known Asia Culture Center, which acts as a focal point for innovation and cross-cultural exchange in Asia.

Gwangju is a city where innovation and history converge to create a vibrant atmosphere that is both hopeful about Korea's future and reflective of its past. In addition to being an important part of South Korea's national landscape, its economic might, cultural significance, and dedication to sustainable development make it a city with a global perspective. Gwangju is a living example of the resilient nature of its people and their ability to think creatively in the face of adversity.

6.2 Background and Origins

Diverse caregiving policies have been introduced by the central government to support vulnerable populations. However, these policies provide a limited range of services to specific groups, primarily focusing on the elderly, people with disabilities, and children. As a result, many residents who require assistance but do not fall into these categories are left without access to essential caregiving services, exacerbating inequality in the use of public resources. The lack of inclusivity in these policies creates disparities, leaving out individuals such as younger adults with chronic illnesses, caregivers themselves who require respite support, or low-income individuals who may not meet strict eligibility criteria.

The "Long-Term Care Insurance for the Elderly," a representative policy created to offer care services for senior citizens, is one well-known example. This policy is very selective, though, as it requires applicants to have a minimum of six months' worth of documented medical treatment history. Approximately 22.1% of elderly people are denied care because of this strict requirement, even though it is clear that they need help. Access to care is further hampered by flaws in the system that provides services. Finding and accessing essential services can be extremely inconvenient for many citizens who struggle to navigate the intricate and disjointed system. This inefficiency emphasizes how urgently more inclusive, approachable, and all-encompassing caregiving policies are needed.

6.2.1 Key Innovative Aspects

This project introduces key innovations in caregiving services to ensure accessibility, efficiency, and inclusivity for all citizens in need. Unlike traditional care models that limit eligibility to specific groups, this initiative allows any citizen to access care services whenever needed. Care Call has a simplified application process that allows people to apply only once for ongoing assistance. Additionally, the system guarantees required visits for those classified as high-risk in order to safeguard vulnerable individuals who might have trouble applying themselves. Each service user is paired with a dedicated citizen-care manager through a one-to-one matching system, guaranteeing individualized and reliable assistance. With the support of a substantial budget to improve availability and efficacy, the project also broadens the scope of caregiving by launching 16 new services. A specialized division has been established to oversee the administration and delivery of services, with the purpose of strengthening its implementation.

Extensive multi-sector collaboration was used to develop the project, combining resources and expertise from the academic, political, private, and public domains. A 17-member task force engaged in 26 rounds of intensive 10-h discussions, incorporating insights from scholars, civil servants, caregivers, welfare organizations, and civic groups to shape a well-rounded, effective caregiving system. Business agreements were signed with 13 organizations, including a medical association, a social welfare association, and a health insurance provider, to guarantee smooth service delivery. This made it easier to identify service users and coordinate care delivery. In order to ensure sustainability and adaptability in the changing caregiving landscape, the project underwent a structural reorganization concurrently, rearranging pertinent teams and institutions to create a more creative and agile execution framework.

6.2.2 Desired Outcome and Measurement

6.2.2.1 Desired Outcome

By offering comprehensive care services that support social inclusion and well-being, the project seeks to guarantee that individuals of all ages can lead healthy and satisfying lives. By providing basic necessities like food, health, and mental health, the program aims to prevent cases of lonely death, depression, and suicide by building robust social support systems. In order to ensure that no one is left behind, the system incorporates a proactive approach whereby Dong-office care managers actively check on people who might need assistance, even if they haven't applied. The one-time application process ensures a quick, easy, and user-friendly experience for citizens who do apply, eliminating administrative obstacles and increasing service accessibility.

The project's main objective is to end care poverty by providing low-income people with services that guarantee them receive quality, fair care. By offering varying degrees of support according to the severity of an individual's needs rather than their financial situation, economic inequality is addressed and care services become equitable and needs-based. Funding from the local government is used to develop new care services, boost the care service market, and generate employment opportunities in order to maintain and grow these initiatives. Furthermore, by encouraging close coordination between the public, private, and social sectors, the project helps important stakeholders create a caregiving system that is more inclusive, resilient, and sustainable (Fig. 6.1).

6.2.2.2 Measurement

The 'Gwangjuness Community Care' project in Gwangju, Republic of Korea, has implemented a comprehensive measurement system to evaluate its effectiveness and ensure that care services are delivered equitably and efficiently. Monitoring service accessibility and utilization is a crucial component of this system, guaranteeing that people of all ages and socioeconomic backgrounds get the assistance they require. In order to better address care gaps, policymakers can continuously improve the program by monitoring the number of beneficiaries, service types, and frequency of

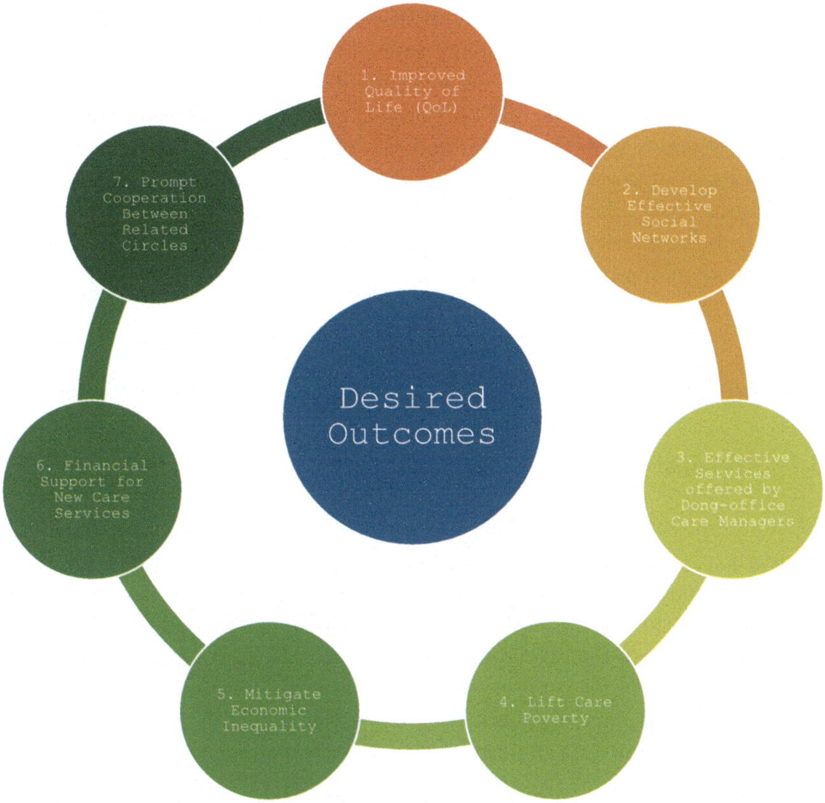

Fig. 6.1 Summary of desired outcomes of the project

use. Care managers proactively reach out to check on vulnerable individuals, paying particular attention to identifying those who might not actively seek help. Surveys of user satisfaction and direct feedback systems are also used to gauge the project's success, enabling modifications that enhance the user experience and service quality.

Data is gathered on important indicators like nutritional status, mental health outcomes, and decreases in avoidable health crises like depression, suicide, and lonely deaths in order to evaluate the project's effects on social and health well-being. The program can assess gains in community involvement, personal well-being, and general quality of life by combining social and health data. Additionally, by guaranteeing that low-income people receive quality care services without incurring financial burdens, economic measures monitor the project's impact on lowering care poverty and mitigating inequality. To guarantee equitable and balanced resource distribution, the efficacy of the graduated support system—which offers aid based on need rather than income level—is routinely assessed.

In order to assess the overall impact on Gwangju's care service market as well as the financial viability of new care services, the project also uses structural and economic performance evaluations. With an emphasis on how the project promotes employment opportunities in caregiving roles, local government funding is closely monitored to evaluate the return on investment in social care and job creation. Efficiency, service integration, and cooperative governance are the criteria used to evaluate partnerships with the public and private sectors, guaranteeing that cooperation results in a more robust and resilient care infrastructure. Through the use of an evidence-based, data-driven methodology, "Gwangjuness Community Care" continues to be responsive and flexible, changing constantly to accommodate the various and expanding needs of its residents.

6.2.3 Implementation

6.2.3.1 Principal Parties and Partners

The 'Gwangjuness Community Care' project operates through a multi-level governance structure, ensuring seamless collaboration between public institutions, private organizations, and local government offices. At the city level, officials create operational guidelines, supervise collaborative governance, and launch Care Call, a one-call application system that expedites service access. While citizen outreach initiatives raise awareness of the support options available, a specialized computer system has been created to manage applications and monitor the use of care services. In order to improve service delivery, the city is also essential in identifying best practices and training care managers. Operating 16 recently created care services, overseeing service provider selection, quality assurance, and financial management are all under the purview of the Gu-office. Additionally, it helps Dong-office care managers, who are frontline staff who make sure to interact directly with citizens who are in need.

Care managers perform both application-based and required home visits at the Dong-office level. These visits are followed by care-level evaluations, personalized care plans, service referrals, and ongoing follow-ups to guarantee the efficacy of the services. One important public organization, the Gwangju Metropolitan Social Service Institute, handles emergency care services and caregiver education, meeting pressing community needs. In the meantime, private organizations help provide services in 15 specialized care areas, utilizing knowledge from fields such as architecture, medicine, and welfare to build comprehensive support networks. In order to maintain and grow the initiative, the city assembly is essential in passing legislation and approving spending plans. The program's capacity to successfully reach vulnerable populations is further strengthened by the cooperation of partner organizations, such as social welfare associations, medical societies, and health insurance companies, which work together to identify citizens in need and provide extra resources.

6.2.3.2 Resources for Implementation

The "Gwangjuness Community Care" project has strong institutional and financial support for its execution. The government's strong commitment to sustainable and equitable caregiving is demonstrated by the $7.74 million the City and Gu-office set aside in 2023 alone to fund care services. In addition to providing direct funding, the project successfully gathers resources from important organizations, guaranteeing that care services are provided to those in need via a network that is well-coordinated. In order to support patients who are discharged and make sure they receive the proper follow-up care, hospitals are essential. Police departments make a contribution by helping victims of crime by giving them access to necessary social services and mental health care. In order to facilitate easier transitions into the proper care programs, the National Health Insurance Service finds and assists those who qualify for senior long-term nursing care.

To find care blind spots and link people with appropriate services, the project also works with social welfare facilities and regional social security councils. These groups are essential in enlisting volunteers from the community and expanding the project's scope beyond programs run by the government. Through the integration of public funding with the infrastructure and expertise of multiple institutions, "Gwangjuness Community Care" guarantees that no person in need is left behind. In addition to increasing the effectiveness of service delivery, this multi-sector collaboration promotes a sustainable care ecosystem in which social welfare organizations, healthcare providers, law enforcement, and government organizations collaborate to make the community safer, healthier, and more inclusive.

6.2.3.3 Tools and Methods Applied

The 'Gwangjuness Community Care' project employs a highly integrated digital system to ensure the swift and efficient delivery of care services. When a Care Call is reported to the Gu-office, it is immediately entered into the computer system to start the process. The system automatically notifies the Dong-office Care Managers, enabling them to react promptly and precisely, rather than depending on phone calls or emails. After being notified, Care Managers go to homes to evaluate the situation and use mobile devices, like tablets, to enter consultation information on the spot. The web-based system instantly synchronizes this data, guaranteeing that all pertinent information is captured and available instantly. Following the evaluation, the required services are chosen and input into the system, which subsequently initiates an automated alert to the relevant private organizations in charge of delivering care.

To ensure smooth coordination, private institutions assign service personnel after being notified, and task details are sent to their smartphone app. Through the app, the designated staff then instantly updates service information and outcomes, enabling Dong-office Care Managers to keep an eye on developments and handle crises as needed. Private institutions use the system to submit cost claims after completing the service, and the Gu-office verifies and processes them for payment. Care Managers follow up with clients after services are completed, either by phone consultations or revisits, and enter evaluations into the system. The project increases efficiency, guarantees accountability, and continuously improves care quality through data-driven evaluation and response mechanisms by incorporating digital tools for real-time monitoring, data entry, and service coordination.

6.3 Obstacles and Solutions

The project faced several challenges during its implementation, particularly regarding institutional coordination and resource allocation. The Gu-office's autonomy in human resource management, which meant that even though it was under the City's jurisdiction, it was not required to follow its instructions, was one of the biggest obstacles. Understanding that hiring enough staff was essential to carrying out the project effectively, the City persisted in negotiating a solution with Gu-office representatives. In order to guarantee the workforce required to support care services at the local level, all Gu-offices created an Integrated Care Division and assigned 368 committed staff members. Budget distribution was another significant barrier, which at first caused the City and Gu-office to be at a standstill in their negotiations. The impasse even garnered unfavorable media coverage, which increased the project's pressure. But after persistent negotiations, the parties came to an amicable funding agreement, settling on a 75% (City) and 25% (Gu-office) split, and obtaining $7.74 million in 2023 to maintain and grow the project.

The skills of frontline care managers, who deal directly with citizens in need, were also crucial to the project's success. In order to make sure that the 1,000 managers—368 from the public sector and 700 from private organizations—completely understood their responsibilities and the project's larger goal, the City arranged 78 intensive training sessions. The scarcity of medical service providers, especially physicians who preferred in-clinic visits over home visits, was another urgent issue. Nonetheless, private organizations reacted forcefully to the 16 recently introduced care services, bolstering the care infrastructure and generating new employment opportunities. Due to the general confidence in oriental medicine, 74 practitioners who specialize in chronic illness and pain management were also included in the project. This creative strategy not only closed the gap in the provision of medical services but also honored regional healthcare preferences, guaranteeing that Gwangju residents would always have access to a wide variety of care options.

6.3.1 Achieved Outcomes

In just seven months (January–July 2023), the project provided 11,983 services to 5,188 people, significantly influencing the well-being of the populace. Through improved health, cleaner living conditions, and more social support, the initiative has enhanced quality of life for people of all ages, from young children to the elderly, by guaranteeing that they can receive timely and appropriate care. Nutritious meal deliveries are now provided to 3,067 beneficiaries who previously struggled to prepare meals because of poverty or illness. Home assistance services are now available to 1,108 citizens who struggled with housekeeping, and 917 people with incurable illnesses can receive home visits from physicians or physical therapists who offer necessary medical examinations and rehabilitation advice. In addition, 2,022 residents now have safer and healthier living conditions thanks to cleaning, pest-proofing, epidemic prevention, and home repair services.

By guaranteeing that 97.8% of low-income service users received full funding, the project has significantly reduced care poverty, thereby reducing care gaps and easing economic inequality. Additionally, the initiative has helped prevent lonely deaths and suicides, especially among elderly people living alone, single-person households dealing with illness, and suicide attempters. The project has simplified access to necessary support by enabling individuals to apply for up to five care services with a single application, making it an effective and user-friendly model of care delivery. The program's sustainability and long-term effects have also been strengthened by the 700 caregiving jobs it has created, which has boosted economic growth. Most significantly, the project has changed how the general public views caregiving by highlighting the fact that it is a shared responsibility rather than an individual burden. This represents a fundamental change in societal attitudes regarding shared support and community well-being.

6.4 Lessons Learned

Publicly managed care services are desperately needed, as the "Gwangjuness Community Care" project has shown, especially in the wake of global crises like COVID-19 and climate change. These difficulties have brought attention to the increased vulnerability of older adults, people with disabilities, and low-income households, who frequently lack access to information about available care services and financial resources. These populations would be more vulnerable to economic hardship, health decline, and isolation in the absence of intervention. In light of this, the project has made care services publicly available, guaranteeing that access to necessary care is determined by actual needs rather than financial resources, in line with the New Urban Agenda's goals for inclusive and equitable cities.

The project has demonstrated how local governments can play a revolutionary role in guaranteeing social sustainability by putting the administration at the forefront of planning and overseeing care services. A strong dedication to creating a resilient and just urban environment is demonstrated by the decision to give care services more funding than any other local policy. The success of the project emphasizes how crucial local efforts are to reaching sustainability objectives, especially when it comes to resolving social injustices and improving community well-being. The idea that a sustainable future must be based on shared responsibility and collective care is further supported by the Gwangju model, which acts as a standard for other cities hoping to institutionalize care as a basic public service.

Related SDGs: Goals 1, 2, 3, 5, 6, 7, 8, 10, 11, and 17.

Acknowledgements We explicitly thank the project owners based in Gwangju, Republic of Korea for their participation and the provision of materials for the 6th Guangzhou International Award for Urban Innovation. We also congratulate them on their achievements in the 2023 cycle of awards.

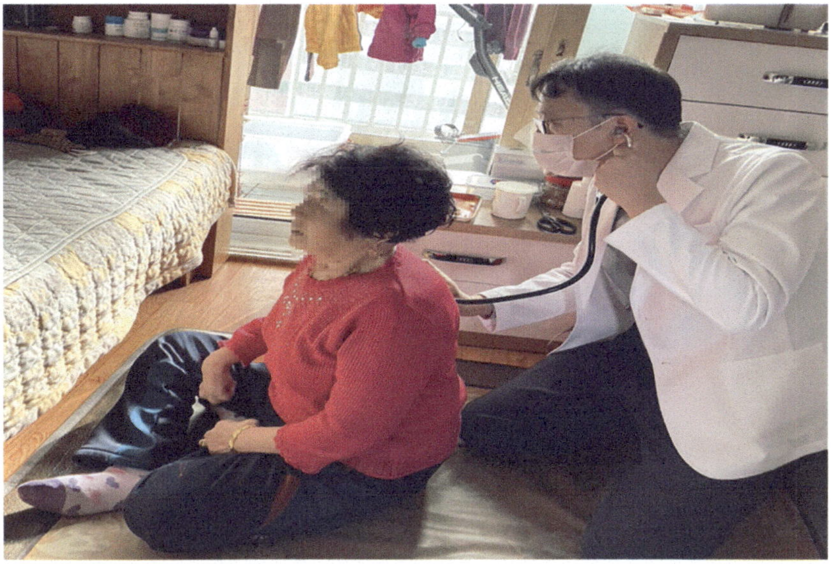

The above images are provided to us by the project owners for their submission.

Chapter 7
UTOPIAS—Units of Transformation and Organization for Social Inclusion and Harmony in Iztapalapa, Mexico

Abstract The UTOPÍAS initiative represents a comprehensive strategy for deep social and urban transformation, designed to overcome the entrenched socio-territorial inequalities that characterize urban peripheries. By focusing on the regeneration of deteriorated public spaces and the construction of a network of large-scale, high-impact public facilities, the project seeks to promote human rights, enhance community well-being, and foster social inclusion. UTOPÍAS revalues public space as a shared community asset, ensuring its equitable distribution while advancing the principles of a 15-min city, where essential services, cultural activities, and recreational opportunities are accessible within a short distance for all residents. Over 540,000 m^2 of formerly deteriorated urban areas have been successfully restored by 12 UTOPÍAS in just four years, turning them into vibrant community centers that now provide weekly services to over 100,000 people. These areas offer wellness, sports, education, and cultural services that directly enhance social cohesion, quality of life, and environmental sustainability. Through the integration of inclusive governance, sustainable urban planning, and public participation, UTOPÍAS has emerged as a potent example of equitable urban development, proving that transformative urban policies can produce resilient, prosperous communities and long-lasting social change. The initiative is related to SDGs 2, 3, 4, 5, 6, 10, 11, 12, 13, 15, 16, and 17.

Keywords Urban Transformation · Public Space Regeneration · Socio-Territorial Inequality · 15-Minute City · Community Well-Being · Equitable Development · Sustainable Urban Planning

7.1 Introduction to Iztapalapa, Mexico

The most populated borough of Mexico City, Iztapalapa, is a vibrant and intricate urban region that is essential to the social, cultural, and economic fabric of the city. It is one of Mexico's most densely populated regions, with 1,835,486 residents and a population density of 16,219.6 persons per square kilometer. Even though Iztapalapa is only 116.67 square kilometers in size, it is a thriving hub for manufacturing, trade,

A. Cheshmehzangi et al., *People-Centered Innovation*, Urban Sustainability,
https://doi.org/10.1007/978-981-96-5230-3_7

and services that greatly boosts the city's economy as a whole. With a GDP per capita of $5,467, the borough demonstrates both its economic difficulties and its tenacity as a major center of business.

Rapid urbanization and migration waves have impacted Iztapalapa's social structure, economic environment, and urban growth throughout its history. The borough still faces urban planning issues, such as the high demand for housing, public services, and infrastructure, even though its annual population growth rate of 1.08% indicates a steady increase in the number of residents. The need for equitable policies to raise everyone's standard of living is highlighted by the GINI Index of 0.39, which shows moderate levels of income inequality.

Iztapalapa is a major commercial and production center in Mexico City, with a thriving economy centered on trade, services, and manufacturing. A strong working-class identity is fostered by the borough's prosperity, which is largely driven by local markets, transportation systems, and industrial zones. Nevertheless, social inequalities still exist in spite of its economic potential, and many locals have restricted access to formal work, healthcare, and education.

Iztapalapa is renowned for its rich cultural heritage and strong sense of community involvement, which go beyond its economic and demographic traits. Every year, thousands of people flock to see Mexico's biggest Passion Play during Holy Week. The goals of community-led programs and urban revitalization initiatives, like UTOPÍAS, are to improve living conditions, address social injustices, and revitalize public areas. Iztapalapa is an intriguing case study in urban transformation and social innovation because of these initiatives, which demonstrate the community's tenacity and dedication to inclusive urban development.

7.2 Background and Origins

Iztapalapa is the second most populated city in the nation, with 1,835,486 residents, or 22% of Mexico City's total population. Because of this enormous population density, there is a greater need for consumer goods, economic opportunities, and basic services. Deep socioeconomic and territorial disparities, however, have long defined the borough and continue to influence its urban environment. About 800,000 people, or 43% of the population, are poor, and 23% work in the unorganized sector, where they face erratic pay and restricted access to social security. Furthermore, 33% of its citizens are in the 15–34 age range, a statistic that highlights the pressing need for community programs, education, and job creation to assist youth. Due to decades of underfunding in its social and physical infrastructure, Iztapalapa has long been regarded as Mexico City's "backyard." Cycles of exclusion and marginalization have been strengthened by a lack of green spaces, a shortage of public spaces, and restricted access to high-quality healthcare and education. In order to guarantee the right to the city and achieve sustainable urban development for its citizens, it is imperative that these disparities be addressed.

To address these pressing issues, the UTOPÍAS project was started as a municipal endeavor with the goal of reversing socio-urban disparities and recovering citizens' rights to a fair, inclusive, and sustainable city. The project incorporates the borough's long-standing social struggles for equitable urban development, environmental justice, and access to public spaces into government policy. Based on the Sustainable Development Goals (SDGs) of the 2030 Agenda and the New Urban Agenda, UTOPÍAS is an ambitious and revolutionary approach to urban planning. By constructing sizable, excellent, multipurpose public areas that foster social cohesion, environmental sustainability, and community well-being, it aims to completely transform urban life in Iztapalapa. The project makes sure that vulnerable groups, especially women and young people, have access to cultural, recreational, and educational opportunities with a focus on gender equality, youth engagement, and human rights. Iztapalapa is showing how systemic change can be sparked by local action through this audacious project, paving the way for a more resilient and equitable urban future.

7.2.1 Key Innovative Aspects

By encouraging shared responsibility between the government, civil society, and local communities, the UTOPÍAS project redefines governance and exemplifies a novel approach to participatory, integrated, transdisciplinary, and trans-sectoral public management. In contrast to conventional urban development models, this transformative approach covers every stage of the project lifecycle, from design and planning to execution, operation, and long-term sustainability. In contrast to traditional top-down governance structures, UTOPÍAS incorporates a variety of stakeholders, guaranteeing inclusive, cooperative, and responsive decision-making processes that take into account the real needs of the people of Iztapalapa.

This shift in governance reflects a qualitative shift in urban policy, reaffirming that the development of public spaces is a social and political commitment to equity, justice, and well-being rather than just a technical or administrative task. UTOPÍAS guarantees that vulnerable and marginalized groups, especially those in poverty, are not only beneficiaries but also actively involved in influencing their urban environment by integrating these ideas into public management. The project's unparalleled scope and ambition are among its most groundbreaking features; it is the first time in Mexico City's history that such a large-scale endeavor has been carried out under the tenets of "first the poor" and "for the poor, the best."

UTOPÍAS has undertaken a fundamental transformation in public management, breaking free from bureaucratic inertia and outdated urban planning models to prioritize rapid, large-scale, and high-impact interventions. The outcomes are evident: more than 540,000 square meters of deteriorated public space have been restored and repurposed in just four years, all with entirely local resources and without the need for outside funding. The idea that major infrastructure projects must rely on outside funding or national-level decision-making is challenged by this effective, economical, and locally driven urban revitalization approach. Furthermore, the project is in line with historical social struggles and global urban trends, especially those that support equitable resource distribution, democratization of public goods, and spatial justice.

The UTOPÍAS model is distinguished by its adherence to the 15-min city concept, which prioritizes accessibility, proximity, and locally focused service delivery. The project's guiding principle is that "for the poor, what is far away does not exist," emphasizing the necessity of placing recreational facilities, cultural venues, and necessary services within walking distance of all locals. Iztapalapa's urban fabric has been reorganized by UTOPÍAS to make it more livable and inclusive by decentralizing urban amenities and guaranteeing that no one is denied access to first-rate public areas. This approach is not just about infrastructure—it is about restoring dignity to communities that have long been neglected and creating a new model for urban governance that prioritizes social cohesion, resilience, and sustainability. Through this initiative, Iztapalapa is proving that bold, locally driven solutions can transform cities, bridging historical inequalities and redefining what is possible in urban development.

7.2.2 Desired Outcome and Measurement

7.2.2.1 Desired Outcome

The main goal of the UTOPÍAS project is to make Iztapalapa a city where everyone's rights are completely upheld and guaranteed by reducing social and territorial inequality. Ensuring the Right to the City, a principle that envisions urban spaces as inclusive, equitable, and accessible for all residents, regardless of their socioeconomic status, is at the core of this initiative. Iztapalapa, a historically underprivileged and marginalized community, has long experienced systemic inequalities in social opportunities, public services, and infrastructure. The project aims to address these disparities and establish an urban setting that promotes community growth, well-being, and dignity by incorporating expansive public facilities, green areas, cultural hubs, and recreation centers.

Throughout the current government administration, the transformation is being implemented gradually, with a final completion date of 2024. Expanding public infrastructure, guaranteeing access to necessary services, and restoring deteriorated areas to become safe, vibrant, and multipurpose urban areas have been UTOPÍAS's top priorities during this time. A city that actively incorporates its most vulnerable residents into its social, economic, and cultural fabric rather than pushing them to the outskirts of urban life is the intended result.

UTOPÍAS is establishing a new standard for urban regeneration by encouraging citizen participation, enhancing local governance, and fortifying social ties. This demonstrates that social justice and sustainable urban development can coexist. By taking these actions, Iztapalapa is not only raising the standard of living for its citizens but also positioning itself as a leader in just urban development.

7.2.2.2 Measurement

A thorough measurement system is being created to monitor both quantitative and qualitative changes in Iztapalapa in order to guarantee the efficacy and impact of the UTOPÍAS project. At the moment, user data is gathered every day and broken down by service providers, sex, age, space usage, and activity participation. This makes it possible to track public participation in the recently created areas and initiatives in real time. The city can determine which services are most in demand, which populations are benefiting the most, and where changes might be required to guarantee fair access for all citizens by examining participation trends.

Although there is a general systematization of data, a more automated and structured database that enables consistent long-term evaluation is becoming more and more necessary. More precise evaluations of the project's effects on urban development, public well-being, and social integration would be possible with the establishment of a centralized platform. In addition to quantitative data, qualitative insights are being obtained through user and visitor interviews and opinion polls, providing insightful commentary on the project's achievements, difficulties, and potential areas for development. These exchanges give UTOPÍAS a clear indication of the needs and goals of the community, guaranteeing that it will continue to be a flexible, people-focused project. Iztapalapa can further solidify its dedication to social and territorial equity by improving these measurement instruments, guaranteeing that the UTOPÍAS project will continue to be a model for inclusive urban transformation.

7.2.3 Implementation

7.2.3.1 Principal Parties and Partners

A multi-stakeholder approach is the foundation of the UTOPÍAS project, which depends on a variety of public, private, and community partnerships to make it successful. Fundamentally, the initiative is run by the Mayor's Office, which handles resource allocation, coordination, and policy direction. However, collaborative governance plays a key role in its execution, uniting different stakeholders through public–private, public-community, and public-public partnerships that enhance the project's sustainability, impact, and reach.

Collaboration with universities and international organizations is a crucial component of the project, especially in the UTOPÍAS's educational spaces. These organizations provide training, research, and scholarly knowledge, guaranteeing that the facilities become centers for professional growth, knowledge exchange, and skill development in addition to being recreational areas. Since many of the project's activities are co-managed with local communities, public-community partnerships are also essential to its success because they encourage residents' long-term commitment and sense of ownership. By enabling local organizations to actively participate in UTOPÍAS's daily operations, maintenance, and programming, these collaborations foster a bottom-up, sustainable approach to urban transformation.

Additionally, public–private partnerships are crucial to the project's execution, especially when working with private companies and non-governmental organizations. By contributing supplies, furniture, and equipment, these organizations have aided the initiative in cutting expenses and increasing service capacity. Furthermore, a significant innovation has been the fusion of science and technology, as demonstrated by the Geological Observatory, which makes scientific research and environmental education accessible to the general public. Another cutting-edge aspect that shows a dedication to sustainable urban development is the application of eco-technologies in facility management and construction. These partnerships have not only contributed resources and expertise but have also benefited from the project by gaining valuable experience in the practical implementation of cutting-edge urban and environmental solutions.

The UTOPÍAS project has been able to optimize its impact through these multi-sector collaborations, transforming the urban landscape of Iztapalapa while promoting social equity, inclusive governance, and environmental sustainability. These collaborations serve as an example of how communities, local governments, and private organizations can collaborate to develop extensive, highly effective urban projects that meet the needs of formerly marginalized groups.

7.2.3.2 Resources for Implementation

The City Congress approved budget allocations for the UTOPÍAS project, which shows a strong government commitment to lowering socio-territorial inequalities in Iztapalapa. The project's financial resources have been carefully managed to guarantee maximum impact, with an emphasis on cost-effective spending and self-administration systems that maximize public monies. In order to maintain the project and meet the district's basic infrastructure and service requirements, the city has created a thorough fiscal plan to increase its own-source revenue. This strategy ensures UTOPÍAS's long-term sustainability by allowing the local government to finance crucial urban regeneration initiatives.

The project has profited from strategic partnerships with academic institutions, government agencies, and private sector organizations in addition to public funding. These partnerships offer significant financial, material, technical, and human resources to improve particular project elements. Public organizations provide technical assistance and policy coordination to guarantee successful implementation, while universities provide research, knowledge, and training initiatives. Partners in the private sector have also contributed by providing supplies, furniture, and equipment, as well as by encouraging the incorporation of eco-technologies into UTOPÍAS's design and administration.

More than 540,000 m^2 of deteriorated public spaces have been transformed into vibrant, excellent, inclusive urban environments thanks to the UTOPÍAS project's ability to maximize its reach and impact through the use of this multi-source resource model. A sustainable, creative governance model is reflected in the efficient distribution and mobilization of resources, showing how local governments can carry out extensive urban transformation initiatives that put equity, accessibility, and community well-being first.

7.2.3.3 Tools and Methods Applied

The UTOPÍAS project is a trailblazing example of participatory, integrated, transdisciplinary, and trans-sectoral public management, fusing community-driven methods with creative governance techniques. With the help of this model, the public sector adopted a new way of working that actively involves different disciplines, sectors, and community organizations in the urban transformation process. The active involvement of local residents is a crucial component of this strategy since it guarantees that the project takes into account the community's actual needs, goals, and desires while simultaneously encouraging a sense of empowerment and ownership among the populace.

In order to create an environment that reflects their shared vision for the future, citizens, including children, were able to directly influence the spaces and services that would ultimately benefit them through the use of participatory design and community planning processes. Popular consultations were held to guarantee the project's success while upholding the native neighborhoods' constitutional right to participate in decisions that impact their land. Through inclusive decision-making processes made possible by these consultations, local opinions were heard and integrated into the UTOPÍAS's final planning and design. In addition to being in line with the values of justice and equity, this inclusive and democratic form of government offered a solid basis for the project's implementation and ongoing operations.

Additionally, local hiring efforts were conducted to find trainers, workshop managers, and construction workers who would help with the actual building of the UTOPÍAS as well as the execution of services and events there. In addition to giving locals opportunities for skill development, the project made sure that the financial gains stayed in the community by employing and training local staff. In addition to creating jobs, this approach strengthened the project's emphasis on capacity building and local empowerment, setting the stage for the spaces to continue operating sustainably long after the original construction phase was finished. These resources and techniques highlight the value of local development, community involvement, and cross-sector cooperation in an inclusive, sustainable, and holistic approach to urban transformation.

7.3 Obstacles and Solutions

During its planning and development phases, the UTOPÍAS project encountered a number of noteworthy challenges, chief among them the opposition to change exhibited by certain local actors and social groups that were rooted in traditional customs and interests. The project's transformative vision was met with resistance from many local stakeholders, particularly those who have historical claims to the public spaces in Iztapalapa. This was primarily due to the fact that the current usages and practices of public areas had become ingrained in the everyday routines and traditions of the local populace. Some people and organizations had long-standing customs or unofficial activities that took up or even misused public spaces, which caused conflict with the UTOPÍAS initiative's main objectives, which were to recover and repurpose these areas for the benefit of the general public and the community.

Alongside this opposition, there were worries about how social groups with specific interests were abusing and influencing public space. Some people were more concerned with preserving the status quo and profiting from unofficial or uncontrolled uses of space than they were with meeting the community's needs as a whole. Tensions resulted from these difficulties because the alteration of public areas endangered the established power structures as well as the unofficial means of subsistence that certain groups had developed there.

In order to overcome these challenges, the UTOPÍAS project put in place a number of mediation procedures designed to promote communication and create agreement between all parties. Constructive dialogues between local actors, citizens, and municipal authorities were greatly aided by mediation. The project adopted inclusive decision-making and citizen participation instead of enforcing top-down solutions, enabling locals to voice their concerns, offer solutions, and influence how the spaces were developed. In order to overcome the initial resistance to change, this participatory approach proved crucial.

In order to involve the community in cooperative conversations where all points of view could be heard and areas of shared interest could be found, assemblies and workshops were planned. These gatherings fostered a sense of collective responsibility for the future of public areas while also assisting in the development of trust between the government and local communities. The project team was able to illustrate the potential advantages of the UTOPÍAS initiative through these open forums, emphasizing how the new areas would offer a range of services and activities that would meet the community's needs as a whole while also providing chances for social and economic advancement.

These efforts ultimately resulted in a number of agreements amongst different stakeholders, balancing personal interests with the project's more general objectives. The result was a varied approach to public space use that accommodated the different social and cultural activities that had previously taken place there while also meeting collective needs. The UTOPÍAS project successfully navigated the obstacles presented by entrenched interests and transformed them into opportunities for greater community participation and empowerment by making sure the transformation was inclusive and flexible enough to fit the local context. This procedure emphasizes how crucial communication, mediation, and inclusive governance are to overcoming opposition and guaranteeing the long-term viability of urban transformation initiatives.

7.3.1 Achieved Outcome

The UTOPÍAS project has transformed Iztapalapa's physical and social landscape with impressive results. More than 540,000 square meters of deteriorated neighborhood public areas have been fully restored during its construction, giving the neighborhood much-needed infrastructure. Twelve sizable, all-inclusive complexes have been built, each providing a range of services centered on coexistence, culture, sports, well-being, and care. The proximity approach and the idea of a 15-min city were used to strategically place these complexes throughout the city, guaranteeing that all citizens could easily access these important areas. These areas have a significant impact; in just two years, over 8 million people have used or visited the UTOPÍAS, demonstrating the initiative's high demand and success. Nearly 100,000 people are served each week. These complexes' breadth and depth of services have promoted both individual and community growth. For instance, there are now twelve swimming pools instead of just one, seven auditoriums instead of three, and eight open forums, all of which offer more space for social events and cultural activities.

Along with fostering social cohesion and cultural enrichment, the initiatives have also resulted in the creation of 148 community orchestras, all of which now receive support from a certified Music School. Furthermore, 27,000 people of all ages have signed up for swimming lessons, which has improved public health and wellbeing. The growth of other educational initiatives, such as the Film & Photography School, which enrolls 2,000 students every four months, has further empowered young people and produced a new generation of talented professionals.

Significant effects on safety and society have also resulted from the project. The crime rate in Iztapalapa has decreased by 57%, indicating that the UTOPÍAS has been successful in creating a safer environment for the locals. The community's cultural and athletic accomplishments have also been transformed by the facilities, as evidenced by Iztapalapa's rise from sixth to third place in the city medal table. These results demonstrate the UTOPÍAS initiative's transformative power by showing how funding public spaces and community-driven services can promote social justice, cultural development, and urban residents' safety.

7.4 Lessons Learned

Six important lessons from the UTOPÍAS project can direct future urban development projects, especially for cities dealing with comparable issues of marginalization, inequality, and subpar public services. It first emphasizes how crucial it is to hear the opinions and experiences of the community. To ensure that social needs are met more precisely, promptly, and efficiently, governments can learn a great deal from the citizens they serve. Involving the community in the project's planning and execution stages has been shown to be a catalyst for tackling regional issues and developing more pertinent and significant solutions.

Second, the project serves as an example of how underprivileged communities in urban peripheries can be guaranteed the Right to the City through the development of comprehensive infrastructure that is centered on human rights, social and spatial justice, equity, quality, gratuity, and proximity. By encouraging a people-centered approach, this strategy makes sure that no one is left behind during the urban development process. For marginalized groups, the layout and arrangement of public areas and services in the UTOPÍAS have helped close the gap and guarantee their access to necessary services. Third, it has been transformative to combine dignified and humanistic care models with exquisite, well-designed, and well-functioning spaces. In addition to providing for the residents' basic needs, these areas foster solidarity, social cohesiveness, and peaceful coexistence, which helps to create thriving and supportive communities.

The project also emphasizes the need for a more thorough and integrated approach to management, projects, and public policy formulation and implementation. The UTOPÍAS have shown that more effective and efficient solutions can be obtained in less time and with fewer resources by adopting a comprehensive approach. Fifth, the UTOPÍAS reiterate how crucial it is to return local governments to their primary role of delivering public services via self-administration. Local governments can free up public resources and guarantee honest and transparent management instead of just administering contracts, opening the door for more significant, large-scale projects like the UTOPÍAS. Last but not least, the population and the government now have a great deal of trust thanks to the successful resolution of long-standing social issues, which has promoted a feeling of intimacy and recognition.

UTOPÍAS's results and lessons have not gone unnoticed. As a result of the city sharing its experiences at international forums and other government agencies adopting similar models, it has gained national and international recognition. Businesses, academic institutions, civil society organizations, and social groups have all recognized the UTOPÍAS and helped make the project a success. As a result, other cities in Mexico and abroad are now implementing the UTOPÍAS model, demonstrating the value of inclusive, participatory urban development and establishing a standard for similar initiatives in the future.

Related SDGs: Goals 2, 3, 4, 5, 6, 10, 11, 12, 13, 15, 16, and 17.

Acknowledgements We explicitly thank the project owners based in Iztapalapa, Mexico, for their participation and the provision of materials for the 6th Guangzhou International Award for Urban Innovation. We also congratulate them on their achievements in the 2023 cycle of awards.

The above images are provided to us by the project owners for their submission.

Chapter 8
Reclaiming Spaces for Children in Public Transportation Through Navi_Station Initiatives in Jakarta, Indonesia

Abstract By improving children's comprehension of station literacy, navigation, public transportation ethics, and safety, the Navi_Station project is a creative effort to reclaim children's space on public transit. The project offers interactive and captivating learning experiences created especially for preschoolers and kids with special needs using an eduplay teaching toolkit. Understanding the value of early exposure to public transportation, this program promotes self-reliance, self-assurance, and inclusivity at an early age. Under the direction of Dr. Eng Widyarini, the project successfully developed a mockup teaching toolkit in 2022 as part of the Short Course on Public Transport in Metropolitan Areas, which sets the stage for future growth. The goal of Navi_Station is to raise awareness of accessibility and safety by incorporating public transportation education into Indonesian schools. In order to improve learning experiences for kids with varying abilities, the project is currently moving forward with the creation of a disability-inclusive toolkit that includes a multisensory book. The Navi_Station project aims to develop a more kid-friendly, accessible, and secure public transportation system for coming generations by focusing on universal design and inclusive mobility. The initiative is related to SDGs 4 and 11.

Keywords Station literacy · Public transport education · Inclusive mobility · Child-friendly transport · Accessibility · Safety awareness · Multisensory learning

8.1 Introduction to Jakarta, Indonesia

Indonesia's capital, Jakarta, is a vibrant megacity that serves as the political, economic, and cultural center of the nation. Jakarta, which has a population of about 10.68 million and is growing at a rate of 1.47%, faces many urban problems, such as traffic, environmental damage, and socioeconomic disparity. The city is one of the most densely populated urban areas in the world, with 15,973 people per square kilometer, covering an area of 661.23 square kilometers. Jakarta, the economic hub of Indonesia, has a GDP per capita of $137,234, which propels the country's growth while simultaneously posing challenges to urban resilience and sustainability.

A. Cheshmehzangi et al., *People-Centered Innovation*, Urban Sustainability,
https://doi.org/10.1007/978-981-96-5230-3_8

Jakarta is leading the way in urban transformation from a sustainability standpoint, but it also faces significant obstacles like air pollution, flooding, poor public transit, and informal settlements. These problems are made worse by climate change, with the city's coastal areas facing a serious threat from rising sea levels. The government has responded by putting in place bold programs like green infrastructure development, sustainable transportation options, and flood mitigation projects. Projects like the Giant Sea Wall (NCICD) and river revitalization initiatives seek to improve climate resilience, while the Jakarta MRT and TransJakarta Bus Rapid Transit (BRT) system have been extended to lessen traffic congestion and carbon emissions.

The Sustainable Development Goals (SDGs) of the UN are in line with Jakarta's sustainability agenda, especially those related to clean water and sanitation (SDG 6), sustainable cities and communities (SDG 11), and climate action (SDG 13). To improve liveability and ecological balance, the city keeps looking into nature-based solutions like urban greening and flood-sensitive spatial planning. Rapid urbanization and socioeconomic inequality, however, pose persistent problems that call for multi-sector cooperation, integrated governance, and creative policies to guarantee Jakarta's sustainability and resilience over the long run.

8.2 Background and Origins

The infrastructure for public transportation in Jakarta is changing quickly. Millions of commuters depend on mass transit services every day thanks to the growth of systems like PT Kereta Api Indonesia (KAI), PT MRT Jakarta, PT LRT Jakarta, PT LRT Jabodebek, and PT KCIC. Preparing the next generation for the use of public transportation in a safe, moral, and effective manner is essential as the city keeps investing in integrated and sustainable transportation solutions. But even with these improvements, many kids still don't know how to use public transportation, especially those in preschool and early elementary school.

In response to this gap, the "Transportation through Navi_Station Initiatives" were created with the goal of educating young users about safety, public transportation ethics, station literacy, and navigation techniques. The program was created using a child-centered approach, acknowledging that early education is essential in forming responsible and self-assured public transportation users. For children with special needs in particular, an engaging and interactive learning approach is necessary to comprehend ticketing systems, train schedules, boarding procedures, and station safety measures.

An eduplay teaching toolkit, a resource created to make learning about public transportation interesting and approachable, was introduced by the initiative to address this issue. The toolkit simplifies difficult transit concepts for kids by utilizing visual aids, practical exercises, and multisensory learning resources. In order to ensure that children acquire the cognitive and behavioral skills necessary to use public transportation safely and effectively, schools, educators, and parents all play a crucial role in fostering this learning process.

Beyond simply promoting train station literacy, the project's larger goal is to promote an inclusive and accessible culture within Jakarta's transportation network. The project is now developing into a disability-inclusive toolkit that includes interactive components and multisensory books to accommodate kids with various learning styles. Navi_Station is establishing the foundation for a kid-friendly, secure, and sustainable urban transportation experience in Jakarta by prioritizing inclusive mobility and universal design.

8.2.1 Key Innovative Aspects

8.2.1.1 Innovations in Navi_Station Initiatives

The Navi_Station project offers a cutting-edge method of teaching young children to use public transportation by combining progressive learning materials, captivating instructional media, active teacher participation, and thorough assessment techniques. This program is notable as an example of how to promote early literacy in public transportation while maintaining accessibility, inclusivity, and involvement.

The progressive content structure of Navi_Station, which takes kids from basic station literacy to more sophisticated navigation skills and safety awareness, is the foundation of its innovation. Preschoolers and kids with special needs can gradually gain confidence in using public transportation thanks to this structured approach. Children are first exposed to basic station components like waiting areas, turnstiles, ticket counters, and signage. As they progress, they gain knowledge of how to navigate, identify platform directions, comprehend train timetables, and behave morally when commuting. In the last stage, which focuses on safety awareness, kids learn about emergency protocols, platform security, and proper public transportation etiquette. A solid fundamental understanding that can be used in actual transit situations is ensured by this methodical learning process.

Navi_Station incorporates interactive and multisensory instructional media to improve student engagement and retention. The project uses bricks, play mats, wipe-and-clean activity books, and paper-craft models in place of traditional classroom-based instruction to give kids a hands-on, immersive experience exploring station environments. By allowing kids to physically engage with station layouts and transit scenarios, these resources not only make learning enjoyable but also support experiential learning. For brand consistency, the project also uses mascots and logos from rail transportation, which aids kids in connecting important visual components to actual public transportation systems. By fostering familiarity, this approach helps young commuters feel less intimidated and more at ease in real station environments.

The Navi_Station initiative's cooperative approach with preschool educators is a crucial component. In order to make sure that the resources are developmentally appropriate and successfully satisfy the learning needs of young children, preschool teachers actively participate in the validation of the curriculum and instructional strategies. Lesson plans, engagement tactics, and activities for varying learning levels are all improved with their input. The project's educational quality is strengthened by this educator-driven validation process, which also makes sure that the content complies with early childhood learning frameworks.

Navi_Station uses a thorough evaluation system that consists of pre-tests, post-tests, and educational visits (edu-visits) to gauge the project's effectiveness. Children take pre-tests to gauge their preliminary knowledge of train station safety and literacy before they interact with the learning materials. Post-tests are used to measure learning outcomes and pinpoint areas that need improvement after educational activities. Furthermore, edu-visits—where kids are seen in actual train stations—offer a chance to evaluate how well they can apply what they've learned in real-world situations. These assessment techniques guarantee that the project has an impact and is flexible, enabling ongoing enhancements to the delivery and content.

The Navi_Station project is revolutionizing how young children in Jakarta interact with public transportation through its creative learning structure, interactive media, educator collaboration, and stringent evaluation process. The project is creating a new generation of self-assured and responsible commuters by making station literacy, navigation, and safety education approachable and entertaining. This will ultimately help create a more sustainable and inclusive urban transit system.

The Navi_Station project's innovative approach to ensuring the smooth integration of station literacy and navigation education is its cooperative approach with important stakeholders in the public transportation and education sectors. To create an integrated wayfinding and signage system at stations, the project actively collaborates with the Jakarta Transportation Agency and significant rail operators, including PT MRT Jakarta, PT LRT Jakarta, and PT Kereta Api Indonesia (KAI). By ensuring that signage, symbols, and directional guides are made accessible and understandable, especially for young children and individuals with special needs, this partnership seeks to establish kid-friendly station environments.

To further strengthen the initiative's long-term effects, attempts are also being made to obtain Ministry of Education support for the inclusion of public transportation literacy in the national school curriculum. By incorporating its eduplay toolkit into early childhood education programs, schools play a critical role in implementing Navi_Station by giving kids a structured learning environment in which to practice station navigation, transport ethics, and safety procedures. Navi_Station guarantees that public transport education becomes a common and sustainable component of Jakarta's urban mobility framework by establishing a bridge between transportation operators, government organizations, and educational institutions. This promotes lifelong learning and responsible commuting practices.

8.2.2 Desired Outcome and Measurement

8.2.2.1 Desired Outcome

By giving young children the necessary station literacy, navigational skills, and safety awareness, the Navi_Station project aims to prepare the next generation of public transportation users. Making sure that kids in DKI Jakarta and beyond are competent and self-assured in utilizing Indonesia's expanding rail transit systems, such as MRT, LRT, and commuter trains, is one of the main goals. By involving kids at a young age, the project hopes to foster moral commuter behavior by highlighting the value of civility, following the law, and acting responsibly in train stations and on transit systems. Furthermore, by making sure that kids and teachers understand station dangers, emergency protocols, and level-crossing safety measures, Navi_Station aims to foster a safety-first mentality. The initiative helps to create a more inclusive, safe, and orderly public transportation culture in Indonesia by encouraging these behaviors at a young age.

The project, which started in 2023 and is expected to conclude by 2027, intends to create and deploy an interactive eduplay toolkit with an emphasis on safety education, navigation, and station literacy. A multisensory book created for kids with special needs is included in this toolkit, which is geared toward kids ages 5–7 and guarantees an inclusive and accessible learning environment. Recognizing the value of multimodal transport education, the project also intends to go beyond rail transport by integrating bus system literacy. In order to guarantee that kids in schools, transit hubs, and community learning centers receive organized and interesting instruction on how to navigate and use public transportation safely and responsibly, Navi_Station hopes to create a standardized and extensively used public transportation education framework by the project's conclusion.

8.2.2.2 Measurement

A thorough evaluation framework has been created to gauge the learning progress and behavioral changes of children who have participated in the Navi_Station project in order to guarantee its efficacy and long-term effects. Three main components make up the structured approach to data collection and assessment: teacher-led assessments using structured questionnaires, direct observation through real-world applications, and pre-test and post-test evaluations.

Improvements in children's comprehension of station literacy, navigation abilities, and awareness of transportation safety are measured using the pre-test and post-test assessments. Children take a pre-test to ascertain their baseline understanding and familiarity with public transportation concepts before using the eduplay toolkit. They take a post-test after finishing the learning activities, which enables the project team to assess the overall efficacy of the instructional materials as well as behavioral changes and knowledge retention.

The project includes direct observation during educational visits to current rail transport modes in addition to theoretical assessments. These edu-visits give kids the chance to use what they've learned in authentic transit settings, like train stations and platforms. The main focus of observations is on how well kids behave as ethical commuters, navigate stations, obey safety instructions, and follow directional signage. This practical evaluation guarantees that learning objectives go beyond activities conducted in the classroom. Preschool teachers who have received the toolkit are also given structured questionnaires created by the Navi_Station team. These surveys enable teachers to assess students' participation, academic development, and real-world application of their knowledge of public transportation. Teachers help identify areas for improvement and adaptation for various learning needs by offering crucial insights into the efficacy of the multisensory learning materials.

Lastly, the Navi_Station team receives all test results and instructor comments for review and development purposes. The toolkit can be continuously improved thanks to this iterative evaluation process, which guarantees that it will continue to be applicable, efficient, and flexible for users in the future. Navi_Station makes sure that its educational framework is impact-driven, scalable, and able to promote long-lasting behavioral change among young people who use public transportation by combining quantitative assessments, evaluations of real-world applications, and educator feedback.

8.2.3 Implementation

8.2.3.1 Principal Parties and Partners

Strong partnerships between the public, private, and community sectors form the foundation of the Navi_Station project, guaranteeing its effective execution, scalability, and long-term viability. Navi_Station establishes a multi-stakeholder ecosystem aimed at promoting rail-based transport literacy for kids in Indonesia by cultivating collaborations with important transport authorities, rail operators, and educational establishments.

In order to incorporate station literacy and public transportation education into Jakarta's changing transit landscape, Navi_Station, a delegation from the Bandung Institute of Technology (ITB), collaborates closely with the city's Transportation Agency and significant national-scale rail operators. PT Kereta Api Indonesia (KAI), PT MRT Jakarta, PT LRT Jakarta, PT LRT Jabodebek, and PT KCIC are some of its most important partners. They all contribute significantly to the provision of educational materials, infrastructure support, and access to actual transit environments for kids' learning experiences.

These public–private partnerships (PPPs) play a key role in making sure that safety procedures, signage, and wayfinding systems complement Navi_Station's educational objectives. Child-friendly transport literacy programs can be seamlessly integrated thanks to the participation of major transportation providers, guaranteeing that the initiative is not merely a stand-alone educational endeavor but rather a component of Indonesia's larger mobility strategy. Additionally, these transportation stakeholders provide knowledge, information, and logistical assistance to enable experiential learning opportunities, like edu-visits to actual train stations, where kids can put their newly learned safety and navigational skills to use.

In order to make transport literacy accessible to young children, Navi_Station prioritizes grassroots engagement in addition to institutional collaborations by collaborating with nearby schools and educational institutions. Sekolah Alam Bogor, an alternative school with an emphasis on experiential learning, is one of its main community partners. Furthermore, the project is being carried out in ten schools throughout DKI Jakarta, catering to students between the ages of five and seven and integrating interactive teaching techniques into the current curriculum.

In order to make sure that Navi_Station's educational resources—such as interactive play mats, multisensory books, and eduplay toolkits—are not only useful in a classroom context but also interesting and inclusive of students with varying learning needs, public-community collaboration is essential. In order to further improve the program, teachers are essential in delivering the content, customizing materials for their students, and offering feedback. Navi_Station creates a sustainable model where public transportation education is incorporated into early childhood education by enlisting the help of local educators, parents, and school administrators.

Through the integration of community-based collaborations and public–private partnerships, Navi_Station builds a solid basis for teaching sustainable transport literacy. The project intends to reach beyond Jakarta with sustained assistance from transportation agencies, rail operators, educational institutions, and community organizations. Eventually, it hopes to impact national education policies and establish station literacy, safety awareness, and navigation skills as essential components of early childhood education throughout Indonesia.

8.2.3.2 Resources for Implementation

The Navi_Station project has been implemented successfully thanks to a combination of in-kind donations, financial support, and strategic partnerships with important stakeholders. Two main funding sources are essential to the project's resource mobilization: the Community Service Program run by the Directorate General of Research and Development Strengthening (DRPTM) of the Ministry of Education, Culture, Research, and Technology, and the ITB Research Program (P2MI), which is sponsored by Institut Teknologi Bandung (ITB). These funding sources give the project the financial support it needs to create learning resources, expand the eduplay toolkit, and guarantee the project's scalability throughout Jakarta's educational institutions.

Along with the monetary donations, Navi_Station also gains a great deal from in-kind support from important public transportation partners. The logos, mascots, and branding materials provided by rail transport and bus operators, such as PT Kereta Api Indonesia (KAI), PT MRT Jakarta, PT LRT Jakarta, PT LRT Jabodebek, and PT KCIC, are incorporated into the instructional content. The learning process is made more interesting and relatable for kids by these visual components, which also serve to strengthen brand recognition and establish a smooth link between the instructional materials and actual public transportation systems.

Furthermore, by providing oversight and direction during the creation of the instructional materials, the Jakarta Transportation Agency plays a critical role in making sure that the project's materials adhere to citywide transportation regulations and continue to be compatible with both present and upcoming transit advancements. This partnership is necessary to guarantee that the Navi_Station project is not only successful in instructing young students but also pertinent to Jakarta's changing urban transportation system.

The combined resources of these different stakeholders guarantee Navi_Station's institutional and financial viability, laying a solid basis for the project's long-term growth and sustainability. The project has the resources required to continue providing young children in Jakarta and beyond with high-quality, accessible, and inclusive public transportation education thanks to public–private collaboration, financial support, in-kind donations, and continuous guidance. This cooperative strategy also creates the framework for upcoming collaborations and financial opportunities, allowing Navi_Station to expand its influence and support the growth of responsible, knowledgeable, and secure public transportation users.

8.2.3.3 Tools and Methods Applied

In order to effectively teach young children about station literacy, navigation, and safety, the Navi_Station project makes use of a range of resources and techniques. This ensures that the content is both interesting and flexible enough to accommodate a range of learning requirements. Since receiving initial funding from the ITB Research Program (P2MI) in March 2023, the project has experienced substantial growth and advancement. This funding made it possible to pilot the project at Sekolah Alam Bogor, a preschool in Bogor, where young students were used to test the educational toolkit. A major step toward reaching a larger student base was taken in July 2023 when the project was able to expand to 10 schools throughout Jakarta thanks to additional funding from the Ministry of Education, Culture, Research, and Technology. The teaching resources were kept current and in line with Jakarta's expanding transportation networks thanks to the partnership with the Jakarta Transportation Agency.

One of the project's main components, the Navi_Station Toolkit, will be made available in November 2023. This toolkit includes a variety of educational resources, including interactive play mats, multisensory books, and safety games, all of which are intended to entertain and educate kids ages five to seven. Train station literacy, the value of safe commuter practices, and using public transit systems are the main topics of the content. Three hundred of the first batch of children will participate in an evaluation to determine the toolkit's efficacy. Pre-test and post-test assessments will be used in this evaluation to measure the children's progress in their comprehension of train station safety, navigation, and general ethical behavior in public transportation environments.

Navi_Station's techniques will change as the project grows in response to input and discoveries from every phase. Next steps for the program include expanding the initiative to other cities outside of Jakarta under the direction of Dr. Widyarini, who started the project. To ensure that the educational materials remain effective and culturally relevant, these expansion efforts will need to be modified to fit the unique circumstances and transportation infrastructure of each city. The toolkit will also be modified in light of each region's particular transportation requirements and cultural viewpoints. To ensure that language, culture, and local transportation norms are accurately reflected in the instructional materials, communities and local organizations from new cities will be encouraged to participate in the toolkit's development.

With each new city offering insightful comments and suggestions to improve the educational experience for kids throughout Indonesia, this iterative process enables the Navi_Station toolkit to continuously improve. Navi_Station will support inclusive, community-driven education while helping to create a national culture of safe and responsible public transportation users by customizing the tools to suit local contexts.

8.3 Obstacles and Solutions

Like any significant educational endeavor, the Navi_Station project has faced a number of obstacles that have needed innovative solutions to guarantee its success and seamless execution. Getting official permissions to use the mascots and logos of different rail-based transportation operators in Indonesia, such as PT Kereta Api Indonesia (KAI), PT MRT Jakarta, and other transport operators, was one of the main challenges encountered. The operators' concerns about intellectual property rights and the commercial use of their branded elements made it difficult to obtain their consent for the use of these logos and mascots, despite the fact that they are essential to the branding consistency of the educational materials.

The project team took a proactive stance to resolve this problem by assuring the operators that the materials were not intended for commercial use but rather were a component of an educational program designed to increase young children's literacy in public transportation. The materials' dual function of educating young children and introducing them to the operators' brands in a non-commercial educational setting was emphasized. The next generation of public transportation users, who may eventually become potential customers, would become more familiar with the brand thanks to this exposure. The project team was able to win over the transport operators by presenting the use of logos and mascots as a tool for early brand engagement rather than commercial advertising.

The project's materials were not incorporated into Indonesia's official preschool and early childhood education curriculum, which presented another difficulty for Navi_Station. It was necessary to establish official channels for the materials' inclusion in educational settings because they were created independently and were not originally included in the national curriculum. The project team applied for program funding from the Ministry of Education, Culture, Research, and Technology in order to address this challenge. This funding was essential for the scaling and expansion of the project into additional schools. Furthermore, efforts were made to register the Navi_Station program with the International Standard Book Number (ISBN) system, a crucial step in making the program materials officially recognized and ensuring that the resources could be easily distributed and adopted by schools across Indonesia.

8.3.1 Achieved Outcome

By making sure that the resources offered are age-appropriate, interesting, and in line with the desired learning outcomes, the Navi_Station project has made great progress toward achieving its educational objectives. Each child can interact with the materials at their own pace thanks to the teaching toolkit's thoughtful design, which takes into account the different skill levels of young children. With content ranging from fundamental station literacy to more complex ideas like navigation skills, ethical behavior, and train safety, the toolkit is especially designed to mirror the educational development stages of kids ages 5–7. This progression makes sure that kids can learn the material efficiently and develop a solid foundation of structured knowledge about public transportation.

Navi_Station has made sure that the teaching toolkit is pertinent to the unique requirements of the Jakarta transportation network by working with the Jakarta Transportation Agency. Children will learn about the trains and stations they will see on a daily basis because the content is especially adapted to the city's rail system. With the toolkit made available to a large number of participating schools, the project's reach has expanded throughout DKI Jakarta. This guarantees that the project is scalable and applicable to the city's larger educational landscape in addition to having an impact in one or two areas.

A great deal of thought went into creating the instructional materials, which include interactive books, props, and entertaining yet instructive hands-on activities. Additionally, the team is preparing these materials for official recognition by submitting the toolkit for inclusion in the International Standard Book Number (ISBN) system. This step is essential to guaranteeing consistency and accessibility in the Navi_Station toolkit's distribution to schools nationwide and its recognition as an official educational resource.

To guarantee efficient implementation and monitoring, the toolkit is currently being distributed in small batches. Before the full-scale distribution starts, the project team can get input, make changes, and enhance the materials thanks to this phased approach. The project's successes in developing a curriculum-aligned resource and working in tandem with the Jakarta Transportation Agency establish Navi_Station as a major contributor to the development of Indonesia's next generation of responsible, knowledgeable public transportation users.

8.4 Lessons Learned

In addition to making great progress in teaching kids about public transit, the Navi_Station project has taught them important lessons, especially about overcoming obstacles brought on by the COVID-19 pandemic. Human mobility has been significantly impacted by the pandemic, as evidenced by the sharp drop in the use of public transportation worldwide, including in Indonesia. As the nation slowly recovers, this change has made it extremely difficult to restore the use of public transportation. There has been a general decrease in public transportation use in the post-pandemic era due to its passivity, especially among younger generations who may have become used to other forms of transportation.

The significance of early education in influencing young people's future behavior, particularly with regard to public transportation, is one of the most important lessons to be drawn from the Navi_Station project. The project intends to instill a sense of familiarity and comfort with public transportation in young children, thereby promoting their acceptance of it as a regular and secure means of transportation. The program indirectly supports the long-term recovery of public transportation usage by raising a generation that is ready and willing to use these services in the future, even though it does not directly address the current ridership decline.

The cooperation with different stakeholders, such as transport operators and the Jakarta Transportation Agency, has brought attention to how crucial multi-sector partnerships are to advancing sustainable urban mobility solutions. Furthermore, the project's adaptable design, which allows it to scale across various cities and cultural contexts, has shown that adaptability is essential when working on projects meant to promote broad behavioral change. Ultimately, the Navi_Station project emphasizes the value of early intervention, teamwork, and cutting-edge teaching resources in creating a sustainable, forward-thinking public transportation culture, particularly in the wake of international emergencies like the COVID-19 pandemic.

Related SDGs: Goals 4 and 11.

Acknowledgements We explicitly thank the project owners based in Jakarta, Indonesia, for their participation and the provision of materials for the 6th Guangzhou International Award for Urban Innovation. We also congratulate them on their achievements in the 2023 cycle of awards.

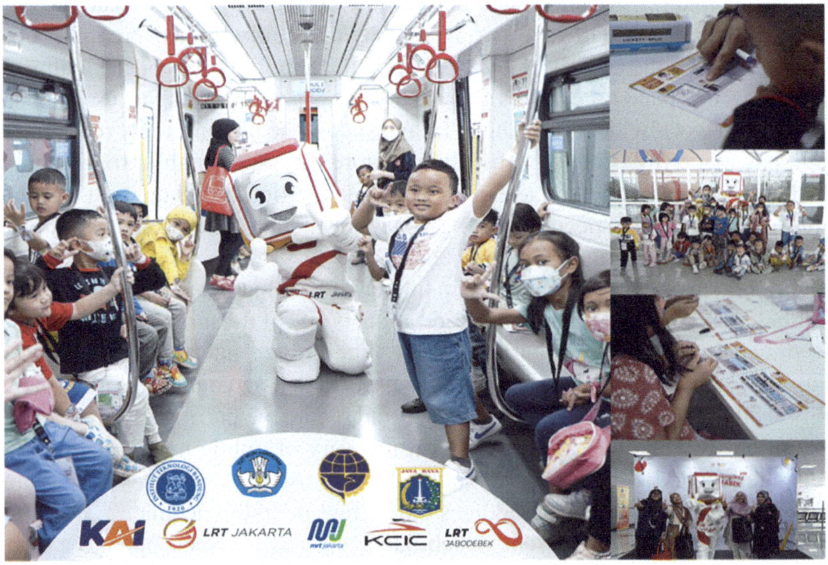

The above images are provided to us by the project owners for their submission.

Chapter 9
An Innovation Towards Pit Emptying Services Through a GIS-Mobile Enabled APP in Kampala, Uganda

Abstract As of 2022, Kampala, the capital of Uganda, had 1,738,000 residents. Because of the city's thriving economy and 5.6% urbanization rate, the population doubles during the day. With a GDP of $2,655, Kampala is a city that is growing quickly and has many sustainability issues. Ensuring equitable services for a healthy population in a clean, livable, and sustainable environment is the responsibility of the Kampala Capital City Authority (KCCA), through its Directorate of Public Health Services and Environment. Water and sanitation services, solid waste management, landfill management, health inspection and education, and sanitation and environmental management are all part of KCCA's core duties, which cover an all-encompassing approach to urban development. By addressing the changing needs of the city's expanding population, these initiatives aim to keep Kampala a clean, healthy, and sustainable place for its citizens. KCCA strives to enhance the quality of life and promote sustainable development in the city's growing urban landscape by placing a high priority on public health and environmental sustainability. The initiative is related to SDGs 1, 2, 3, 8, 11, 12, 13, 14, 15, 16, and 17.

Keywords Kampala · Uganda · Urbanization · Public health · Sanitation · Waste management · Sustainable development

9.1 Introduction to Kampala, Uganda

Uganda's capital, Kampala, is a shining example of urban change and the urgent issues that the Global South's rapidly expanding cities face. Kampala, which has a population of 3,846,102 according to recent estimates, is a center of rapid urbanization, cultural diversity, and economic activity. Its 5.31% annual population growth rate, which is among the highest in Africa, reflects this expansion. Kampala's population density has increased to 9,352 people per square kilometer as more people move there in quest of better economic opportunities, making it more difficult to provide essential services and manage urban resources in a sustainable manner.

A. Cheshmehzangi et al., *People-Centered Innovation*, Urban Sustainability,
https://doi.org/10.1007/978-981-96-5230-3_9

With a surface area of 189 square kilometers, Kampala is comparatively small, but its dense residential areas and expansive informal settlement underscore the difficulties associated with land use and urban planning. The city's infrastructure, such as its waste management, water supply, and transportation system, is severely strained by the dense population. Despite these obstacles, Kampala's strategic location as Uganda's political and economic hub has created a vibrant, if complicated, urban environment.

As the primary driver of prosperity, tourism is a major component of the city's economy. Due to Kampala's rich cultural legacy, thriving marketplaces, and close proximity to important natural landmarks like Lake Victoria and national parks, the city attracts thousands of tourists from both domestic and foreign countries. The need for a more resilient and diversified urban economy is underscored by the fact that this economic reliance on tourism also results in seasonal variations and a reliance on international travel trends. The low GDP per capita of $2,655 in Kampala highlights the need for inclusive growth that offers opportunities for the city's rapidly growing population, especially in non-tourism industries like manufacturing, technology, and green industries.

The fast urbanization of Kampala brings with it both possibilities and difficulties. On the one hand, the expansion of the city promotes economic growth, innovation, and cross-cultural interaction. However, it makes problems with social justice, service accessibility, and environmental sustainability worse. There are now major worries about overcrowding, inadequate housing, and restricted access to clean water and sanitation as a result of the growth of informal settlements, especially in places like Kawempe, Makindye, and Nakawa. The city is under growing pressure to update and expand its infrastructure in order to meet demand as the population keeps growing.

For Kampala to remain sustainable in the future, urban innovation is crucial. By making investments in waste management, infrastructure, and health services, the Kampala government, through organizations like the Kampala Capital City Authority (KCCA), has taken significant steps to address some of these issues. Nonetheless, the city continues to struggle with a lack of reasonably priced housing, poor public transit, and a lack of urban green spaces. As a result, creative solutions are needed to satisfy the needs of the city's expanding population while guaranteeing development that is both environmentally conscious and equitable.

Kampala's sustainability initiatives must consider the necessity of striking a balance between environmental preservation and economic growth. The future of the city depends on issues like pollution, waste management, and climate change adaptation. Kampala's attempts to become a more resilient and livable city may be greatly aided by the creation of green areas, sustainable water management techniques, and clean energy sources. Additionally, Kampala's expanding tech-savvy populace and improved connectivity offer chances to use smart city technologies to creatively and efficiently solve urban problems. Lastly, Kampala's explosive growth presents a singular chance to investigate the relationship between sustainability and urban innovation. As the city develops further, incorporating creative solutions into infrastructure development, urban planning, and environmental management will be essential to building a thriving, sustainable city for the coming generations. The story of Kampala is a useful case study for comprehending the challenges of urban development in quickly expanding cities and the significance of striking a balance between social, economic, and environmental goals in order to achieve long-term urban sustainability.

9.2 Background and Origins

More than 60% of Kampala's population lives in informal settlements, which present serious sanitation issues. Access to adequate sanitation services is still a major concern in light of the country's fast urbanization and 5.31% population growth rate. Since only 10% to 15% of the city is currently connected to a formal sewer system, most citizens must rely on On-Site Sanitation (OSS), which includes septic tanks and pit latrines. This dependence calls for a sustainable and efficient Faecal Sludge Management (FSM) system, but the actual situation is far from ideal. Independent pit emptiers frequently use unsafe and unregulated emptying methods, which leads to the unlawful disposal of waste into open drains, water bodies, and other public areas, thereby increasing the risks to human health and the environment.

In addition to being a vital public service, access to better sanitation is also a critical defense against waterborne illnesses like cholera, typhoid, and diarrhea, which disproportionately impact low-income communities. The Kampala Capital City Authority (KCCA) has determined that FSM is a top priority for improving urban sanitation in response to these urgent sanitation issues. KCCA has worked to innovate by utilizing digital solutions to improve service accessibility, regulation, and accountability after realizing that traditional methods of sanitation service delivery were unsustainable and inefficient.

One of the biggest problems is the restricted availability of regulated toilet emptying services, where awareness, cost, and logistics continue to be major obstacles. Due to their inability to pay for expert emptying services, many locals dispose of their waste improperly, further taxing the urban environment. To address these challenges, KCCA launched the Weyonje Application, an innovative platform designed to bridge the gap between pit latrine users and professional faecal sludge emptiers. The app ensures that sanitation services are provided in a safe, economical, and effective manner by connecting locals with registered and certified emptying service providers.

The launch of the Weyonje App signifies a fundamental change in Kampala's approach to managing and obtaining sanitation services. KCCA seeks to change the urban sanitation scene by combining technology, community involvement, and regulatory supervision, lowering the risks to the environment and public health that come with inadequate FSM practices. Additionally, by encouraging sustainable urban sanitation solutions, the initiative supports Uganda's larger efforts to meet the Sustainable Development Goals (SDGs), especially SDG 6 (Clean Water and Sanitation).

Scalable and creative FSM solutions are becoming more and more necessary as urban populations continue to rise. A model for digitally driven public service delivery, the Weyonje App and other urban sanitation initiatives provide insights that could guide similar projects in other rapidly urbanizing cities in Africa and beyond.

9.2.1 Key Innovative Aspects

An innovative digital tool called the Weyonje client application was created to improve Kampala's fecal sludge management (FSM) and sanitation accessibility. Similar to ride-hailing services like Uber, the app makes it simple for locals to use their phones to request septic tank or pit latrine emptying services. The Weyonje app is changing the way sanitation services are provided in the city by fusing technology, environmental sustainability, and community involvement (Fig. 9.1).

Its digital accessibility is one of its main innovations. The app improves efficiency and service delivery by offering an on-demand platform that makes it easier to find and schedule certified emptying service providers. In addition to guaranteeing prompt waste removal, this technological intervention lowers the risks connected with unhygienic disposal practices, which are prevalent in unregulated FSM systems. The app integrates Village Health Teams (VHTs) and Community Activation Teams to help close the digital divide. These teams make sure that even people living in informal settlements have access to regulated sanitation services by helping low-income residents who do not have smartphones. This strategy encourages inclusive urban sanitation while broadening the app's user base.

Data-driven decision-making is one of Weyonje's key characteristics. By gathering and analyzing data on pit latrine emptying patterns, the app helps the Kampala

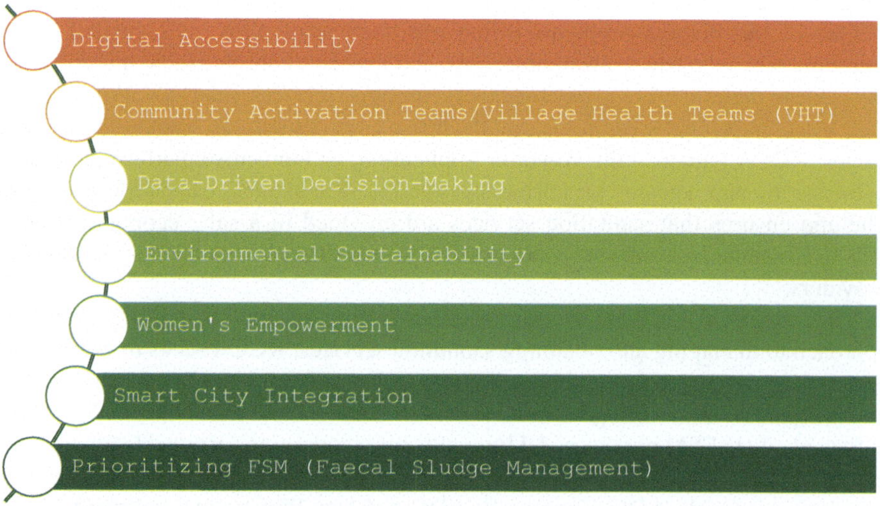

Fig. 9.1 Summary of key innovations introduced by the Weyonje app in Kampala, Uganda

Capital City Authority (KCCA) and other pertinent parties keep an eye on sanitation requirements, make effective use of available resources, and avert public health emergencies. Authorities can lower the prevalence of waterborne illnesses by identifying high-risk areas and implementing prompt interventions by utilizing real-time data.

Sustainability in the environment is central to Weyonje's mission. The app prevents illegal dumping, which is a major source of water contamination and land degradation in Kampala, by making sure that fecal sludge is transported and treated at facilities that are authorized. This controlled strategy lessens the detrimental environmental effects of subpar FSM practices and is in line with the city's larger objectives for sustainability and climate resilience. Along with promoting women's empowerment, the Weyonje initiative supports SDG 5 (Gender Equality). Although FSM has historically been a male-dominated industry, the app encourages women to participate in emptying services, opening up economic opportunities for them. This change fosters a more equitable workforce by encouraging gender-inclusive employment in urban sanitation.

Kampala's Smart City Integration includes the Weyonje app, which helps to update urban services. It supports Uganda's urban transformation agenda by digitizing sanitation management, demonstrating how smart technologies can enhance the provision of public services. Furthermore, Kampala's dedication to FSM as a top priority demonstrates the city's proactive approach to addressing sanitation issues and establishes a standard for other metropolitan areas looking for long-term fixes. Weyonje is transforming sanitation in Kampala by fusing technology, community involvement, and sustainability. They are providing a scalable model that may serve as an inspiration for similar projects in other quickly urbanizing cities.

9.2.2 Desired Outcome and Measurement

9.2.2.1 Desired Outcome

The goal of the Weyonje initiative is to revolutionize Kampala's sanitation industry in both the short and long term. The project's short-term goal is to modernize private sector sanitation service providers' business practices, especially within the next year. Nowadays, a lot of operators still use antiquated techniques, like paper-based records, which are prone to being misplaced, inconsistent, or nonexistent in certain situations. The initiative aims to create a standardized, dependable, and user-friendly system for monitoring emptying services, payments, and customer interactions by implementing digital documentation via the Weyonje app. This change will increase the sector's financial accountability, boost customer satisfaction, and improve operational efficiency.

The initiative seeks to develop a more data-informed regulatory framework in addition to increasing business efficiency. Weyonje will offer real-time insights into sanitation demand, service frequency, and coverage gaps by gathering and centralizing service data. By identifying underserved areas, allocating resources optimally, and proactively addressing sanitation needs, this evidence-based approach will help both public and private operators improve the health of Kampala's citizens.

In the long term, Weyonje aspires to have an impact on legislative changes that require institutional departments to make decisions based on data. In the past, estimates or out-of-date reports have served as the foundation for sanitation policies and resource allocations, which has resulted in inefficiencies and subpar service delivery. Policymakers will have access to precise, fact-based data to guide planning, budgeting, and service enhancements by incorporating Weyonje's real-time FSM data into urban governance. The transition to open, data-supported reporting will guarantee that sanitation management improves in sustainability, efficiency, and responsiveness to the city's expanding urban population.

The degree of institutional adoption and regulatory support will also be used to gauge the initiative's long-term success. The goal of the project is to create a sustainable framework where data-driven sanitation services are institutionalized, financed, and continuously improved by integrating FSM tracking and reporting into the larger urban management strategies of the Kampala Capital City Authority (KCCA). In the end, Weyonje is a catalyst for systemic change rather than merely providing sanitation services. The project lays the groundwork for a healthier, cleaner, and more resilient Kampala in the years to come by modernizing private sector operations, encouraging evidence-based governance, and supporting sustainable sanitation policies.

9.2.2.2 Measurement

A strong measurement and evaluation system has been put in place to guarantee the Weyonje initiative's efficacy and sustainability. To evaluate progress, impact, and areas for improvement, this system gathers, processes, and analyzes important sanitation data. Customer-specific information like name, gender, phone number, and type of facility emptied (septic tank or pit latrine) are among the data gathered. The customer's geographic location is also noted, along with the division and parish where the service is rendered. This location-based data is essential for mapping underserved areas, spotting trends in service demand, and making sure sanitation services are dispersed fairly throughout Kampala's urban landscape.

On the side of the service provider, information is gathered about the provider's name, contact details, service type (gulper operator or cesspool emptier), and service delivery date. The precise location of sludge disposal is also noted in order to track sustainability and compliance. This ensures that all waste is appropriately disposed of at approved treatment facilities, lowering illegal dumping and environmental pollution. The initiative's real-time tracking of the volume of sludge deposited at the assigned disposal sites is a noteworthy innovation. The initiative can evaluate capacity limitations, optimize waste treatment facility operations, and predict future infrastructure needs by methodically documenting the amount of sludge managed. A group of knowledgeable data scientists and biostatisticians carefully examines this abundance of data, allowing them to:

- Better Urban Planning: Policymakers can prioritize investments in sanitation infrastructure by identifying gaps in sanitation services.
- Effective Resource Allocation: Authorities and service providers can allocate resources where they are most needed thanks to evidence-based insights.
- Performance Monitoring: To guarantee ongoing progress, the program keeps tabs on customer satisfaction, provider effectiveness, and service frequency.
- Public health surveillance: The information makes it possible to identify disease outbreaks linked to poor sanitation early and take prompt action.

Weyonje is revolutionizing Kampala's sanitation scene by utilizing data-driven decision-making to make sure that each intervention is impactful, quantifiable, and targeted.

9.2.3 Implementation

9.2.3.1 Principal Parties and Partners

A multi-stakeholder approach that brings together public institutions, private sector players, community organizations, and technical experts is essential to the Weyonje initiative's successful implementation in Kampala. A public–private partnership (PPP) model, which guarantees effective and sustainable sanitation service delivery, is at the heart of the initiative. The Kampala Capital City Authority (KCCA) is responsible for overseeing and regulating licensed toilet emptying service providers to make sure they follow public health and environmental regulations. In order to maintain sanitation compliance throughout the city, the KCCA also provides vital infrastructure, such as designated dumping bays and treatment facilities.

One of the main actors in carrying out on-the-ground operations is the private sector, which includes authorized cesspool and gulper service providers. These service providers can effectively manage job orders, monitor service delivery, and keep open and transparent business records thanks to the Weyonje app, which is a crucial digital tool. Ongoing capacity-building and digital training initiatives are carried out to guarantee efficient app use, giving service providers the tools they need to boost their operational effectiveness. Village Health Teams (VHTs) and Community Activation Teams (CASTs) assist in providing sanitation services to underserved and low-income communities in addition to sanitation operators. By bridging the digital divide, these teams make sure that households can still request sanitation services even if they don't have smartphones or internet access.

Data scientists and biostatisticians are essential in examining service coverage, customer trends, and environmental impacts as part of the initiative's data-driven decision-making framework. Their knowledge facilitates evidence-based planning, which helps KCCA and legislators improve sanitation tactics and distribute resources effectively. The impact and sustainability of the initiative are increased through additional cooperation with international development agencies, non-governmental organizations (NGOs), and government regulatory bodies. These organizations advocate for better sanitation policies, provide financial support, and offer technical expertise. Environmental organizations also make sure that waste management procedures are in line with sustainability objectives, reducing the dangers of unlawful dumping and groundwater contamination.

Lastly, the initiative is actively improved and engaged with by local communities, residents, and customers. Their involvement affects long-term sanitation planning and service improvements in addition to promoting adoption and behavioral change. The Weyonje initiative is establishing a standard for creative, inclusive, and sustainable urban sanitation solutions by incorporating it into Kampala's larger Smart City framework.

9.2.3.2 Resources for Implementation

In order to ensure sustainable sanitation service delivery in Kampala, the Weyonje initiative must be implemented successfully, which calls for a thorough distribution of resources across multiple domains. The efficacy and long-term viability of the project depend on these resources, which can be divided into financial, human, technological, regulatory, community, and analytical components.

Financial Resources and Infrastructure

The Weyonje digital platform's development and upkeep depend heavily on adequate funding, which guarantees that the app's usability, security, and accessibility are continuously enhanced. Infrastructure development also requires funding, especially for sludge treatment plants, designated dumping bays, and sanitation logistics—all of which are critical for the safe disposal of waste. Investing in low-income households' subsidies or incentives also improves service accessibility, guaranteeing that the program benefits vulnerable communities.

Human Capital and Training

Implementing and maintaining the initiative depends on having a competent workforce. Public health experts to supervise regulatory compliance, service providers for sanitation operations, and app developers and IT specialists for platform management are all included in this. For toilet-emptying service providers, ongoing training programs guarantee effectiveness, adherence to safety regulations, and appropriate app use. Additionally, Village Health Teams (VHTs) and community leaders can facilitate service requests in digitally underserved communities thanks to capacity-building initiatives.

Technological and Data Management Tools

Digital tools are essential to the initiative's data collection, tracking, and assessment. In order to facilitate real-time decision-making, the Weyonjc app keeps track of service requests, customer demographics, and disposal trends. Remote sensing and GIS mapping technologies also improve spatial analysis, allowing authorities to pinpoint high-risk areas for focused interventions. To preserve regulatory compliance and safeguard user privacy, a strong data security system is required.

Regulatory and Legal Compliance Resources

Enforcing sanitation policies through robust regulatory frameworks is essential to Weyonje's success. Legal resources prevent unregulated dumping by ensuring that all service providers adhere to public health regulations. Working together with government representatives guarantees that licenses are kept current, accountability is maintained, and sanitation regulations are enforced.

Community Engagement and Awareness

Participating in local communities is essential for boosting adoption and promoting sustained behavioral change. Funds are required for targeted education initiatives, workshops, and public awareness campaigns that emphasize the value of proper sanitation and waste management. Collaborations with community organizations, NGOs, and media outlets promote involvement and aid in the dissemination of important information.

Monitoring, Evaluation, and Sustainability

Continuous monitoring and evaluation (M&E) is necessary to determine Weyonje's impact. Data-driven policy changes and efficient service delivery are guaranteed by specialized resources for data analysis, field research, and performance monitoring. By establishing key performance indicators (KPIs), the model can be replicated in other urban centers and improved continuously.

By mobilizing these diverse resources, the Weyonje initiative is well-positioned to drive impactful, technology-driven sanitation solutions in Kampala, aligning with public health, environmental sustainability, and smart city development goals.

9.2.3.3 Tools and Methods Applied

The development and implementation of the Weyonje app rely on a comprehensive suite of tools and methods to ensure efficient faecal sludge management, user engagement, and regulatory compliance. These tools encompass technological, analytical, and operational components, all of which contribute to the seamless functionality, accessibility, and sustainability of the initiative (Fig. 9.2).

Modern mobile application development frameworks like Android Studio, Flutter, and React Native were used in the creation of the Weyonje app, guaranteeing accessibility and cross-platform compatibility for a diverse user base. Customer information,

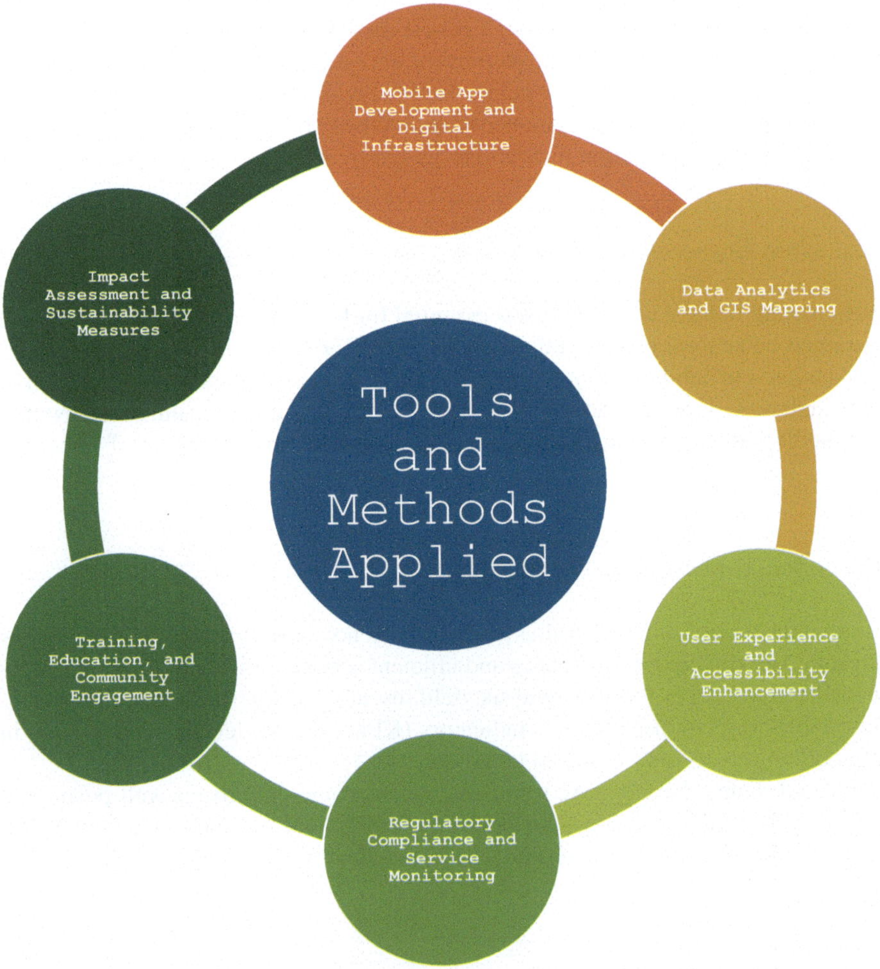

Fig. 9.2 Summary of a comprehensive suite of tools and methods to ensure efficient faecal sludge management, user engagement, and regulatory compliance

operational data, and sanitation service records can all be safely stored and retrieved with the help of database management systems like PostgreSQL and Firebase. Scalability is a feature of cloud computing services like Google Cloud and Amazon Web Services (AWS) that guarantees the system can accommodate growing user demand without sacrificing performance.

The Weyonje initiative relies heavily on data-driven decision-making. The app tracks and analyzes important sanitation trends using sophisticated analytics and visualization tools like Tableau and Power BI. The app also incorporates Geographic Information Systems (GIS) to map the coverage of sanitation services, pinpoint areas with high demand, and help service providers plan their routes more efficiently. In underprivileged areas, these location-based insights enable targeted interventions and real-time monitoring.

To guarantee a smooth experience for both clients and service providers, the Weyonje app is made with user-friendly interfaces using UI/UX design tools like Figma and Adobe XD. Customer ratings and in-app surveys are examples of feedback mechanisms that allow for ongoing enhancements based on user feedback. Furthermore, multilingual support guarantees inclusivity by enabling users who do not speak English to efficiently use the platform.

To align with national sanitation policies, the app integrates regulatory compliance tools that allow Kampala Capital City Authority (KCCA) officials to oversee sanitation operations and enforce waste disposal guidelines. Automated reporting mechanisms ensure compliance with environmental standards by tracking the volume of faecal sludge removed and its proper disposal at designated treatment sites. Smart sensors and IoT devices are being explored to further enhance real-time monitoring of sanitation infrastructure.

To teach service providers and community health teams how to use the Weyonje app correctly, a mix of online and in-person training techniques is used. Both tech-savvy and non-tech-savvy users can interact with the platform efficiently thanks to training materials that include interactive webinars, video tutorials, and on-site workshops.

Monitoring and evaluation (M&E) frameworks serve as a guide for the implementation process, tracking the impact on public health, user engagement, and service efficiency. Response times, service completion rates, and the effectiveness of waste disposal are examples of key performance indicators (KPIs) that are routinely measured. The results of these evaluations are utilized to improve the app's features and expand its reach into more cities. The Weyonje initiative guarantees sustainable, effective, and data-driven fecal sludge management by utilizing these various tools and techniques, which is in line with Kampala's larger urban innovation and smart city objectives.

9.3 Obstacles and Solutions

To guarantee its success and uptake, the Weyonje app's implementation in Kampala had to overcome a number of obstacles. Resistance to change among service providers, many of whom were used to manual, traditional methods of operation, was one of the main challenges. In order to solve this, in-depth training sessions were held, giving them the knowledge and self-assurance they needed to switch to digital record-keeping and service management. Their worries were allayed by practical assistance and ongoing involvement, which illustrated the real advantages of increased productivity and corporate transparency.

The digital divide was another major issue, especially in informal settlements where access to smartphones is restricted. Partnerships with neighborhood organizations made it possible to provide reasonably priced smartphones in order to close this gap, and the app's user-friendly interface featured options for local languages. This made it possible for people who were not familiar with digital technology to use the platform efficiently. Another obstacle was money, especially when it came to getting the money required for app development, upkeep, and growth. Working together with NGOs, international development organizations, and corporate sponsors was necessary to address this issue. These collaborations ensured the initiative's long-term viability by contributing technical know-how in addition to financial support.

Another important element in the app's success was community engagement. At first, many locals had doubts about the advantages of receiving sanitation services via a digital platform. Large-scale awareness campaigns were started with the help of Activation Teams, local influencers, and community leaders in order to foster trust and promote broad adoption. These groups were crucial in answering queries, outlining the advantages of the app, and showing how it could enhance neighborhood sanitation and public health. In order to overcome these obstacles and guarantee its beneficial effects on Kampala's sanitation infrastructure and general public health, the Weyonje initiative combined education, funding, community engagement, and technological adaptation.

9.3.1 Achieved Outcome

The Weyonje project has greatly increased the effectiveness and accessibility of fecal sludge management, resulting in notable advancements in sanitation services throughout Kampala. Service providers can now function with increased accountability and transparency by incorporating digital solutions, guaranteeing that waste disposal complies with legal requirements and reducing environmental risks. Fair interventions for the urban poor, especially in informal settlements where access to sanitary facilities is still difficult to come by, have also been a focus of the initiative. It has raised the safe management of fecal sludge to 60% through targeted subsidies and relief programs, offering crucial services to communities that were already at risk, many of which experienced extreme financial hardships as a result of the COVID-19 pandemic.

The initiative has been successful in changing sanitation practices, as evidenced by the positive behavioral changes seen in over 15,509 people. Alongside this shift, public–private partnerships are becoming stronger, creating a cooperative atmosphere where government organizations, private businesses, and neighborhood associations collaborate to develop sustainable sanitation solutions. Additionally, the project has been successful in securing more funding cycles from the Bill & Melinda Gates Foundation, guaranteeing its ongoing growth and influence.

The opening of a medical call center, which offers crucial healthcare information and assistance and further integrates public health services with sanitation management, is another significant milestone. Authorities can provide early warnings for regions in need of more resources by using evidence-based data, which enables proactive planning and intervention. Government commitment to enhancing sanitation policies and infrastructure has been strengthened through partnerships with the Ministry of Water and Environment and the Ministry of Health.

A key factor in the initiative's success has been empowering local communities to spearhead sanitation initiatives, resulting in a community-led sanitation model that guarantees locally driven and sustainable solutions. Additionally, the program has aided in economic empowerment, job creation, and local entrepreneurship by assisting small businesses that provide sanitation services. In addition to helping Kampala, this model establishes a standard for other cities looking to combine economic growth and sanitary advancements.

The initiative's fecal sludge collection efficiency has significantly increased from 44 to 61% from an operational standpoint. As more waste is being safely transported to approved treatment facilities, this progress has led to a quantifiable decrease in the environmental impact. Efficiency, effectiveness, accountability, and transparency have all been further improved by digital monitoring of service providers; an astounding 721,141,800 m^3 of waste has been safely disposed of, and over 15,509 jobs have been successfully finished. These accomplishments demonstrate how technology, teamwork, and community involvement can propel sustainable urban development and highlight the initiative's revolutionary effect on Kampala's sanitation landscape.

9.4 Lessons Learned

The importance of data-driven decision-making in managing urban sanitation has been highlighted by the Weyonje initiative. Policymakers and service providers have successfully targeted interventions by utilizing evidence-based insights, especially in informal settlements where sanitation needs are most pressing. In addition to enhancing planning, having access to precise, up-to-date data has made it possible to allocate resources more fairly and guarantee that vulnerable groups get the assistance they require.

The significance of solid public–private partnerships in providing necessary services has been another important lesson learned. Through encouraging cooperation between public and private organizations, the program has made sanitation services more accessible while also enabling business owners to legally establish their enterprises. In addition to improving record-keeping and operational efficiency, this strategy has helped create a more sustainable urban sanitation ecosystem. The model's success demonstrates its scalability and reproducibility, providing insightful information for cities around the world dealing with comparable issues.

The initiative has also highlighted how transformative solutions can be driven by group decision-making. By involving a variety of stakeholders, including local communities, service providers, legislators, and regulatory agencies, sanitation interventions have been made inclusive, workable, and generally accepted. Maintaining these initiatives has been made possible in large part by the integration of legal and regulatory frameworks, which has allowed Kampala to embed best practices for sanitation within preexisting governance structures and negotiate complicated policy environments.

Another important lesson has been to invest in skill development. Sanitation workers now have the resources they need to use the Weyonje app efficiently thanks to capacity-building initiatives that emphasize digital literacy and data management. In addition to enhancing service delivery, this workforce development strategy has given employees new chances for career advancement. Additionally, with intensive training and support, the challenge of overcoming resistance to change has been successfully addressed. With the correct direction, many service providers who were first reluctant to embrace the technology have made the seamless transition to the new digital system, realizing its advantages for operational effectiveness and business expansion.

The necessity of addressing the digital divide in urban sanitation initiatives is the last but no less important lesson. It is crucial to create solutions that are easy to use and available to everyone because many of the most marginalized groups do not have access to smartphones or digital literacy. The initiative has shown that technology-driven interventions can be inclusive and equitable through its efforts to provide reasonably priced smartphones, guarantee multilingual support, and involve community activation teams. In the end, the Weyonje project has demonstrated that finding sustainable sanitation solutions calls for a multipronged strategy that incorporates technology, data, governance, and community involvement. These lessons offer a path forward for other cities looking to update their sanitation systems, demonstrating that major advancements in urban sustainability and public health can be made with the correct mix of creativity, cooperation, and inclusivity.

Related SDGs: Goals 1, 2, 3, 8, 11, 12, 13, 14, 15, 16, and 17.

Acknowledgements We explicitly thank the project owners based in Kampala, Uganda, for their participation and the provision of materials for the 6th Guangzhou International Award for Urban Innovation. We also congratulate them on their achievements in the 2023 cycle of awards.

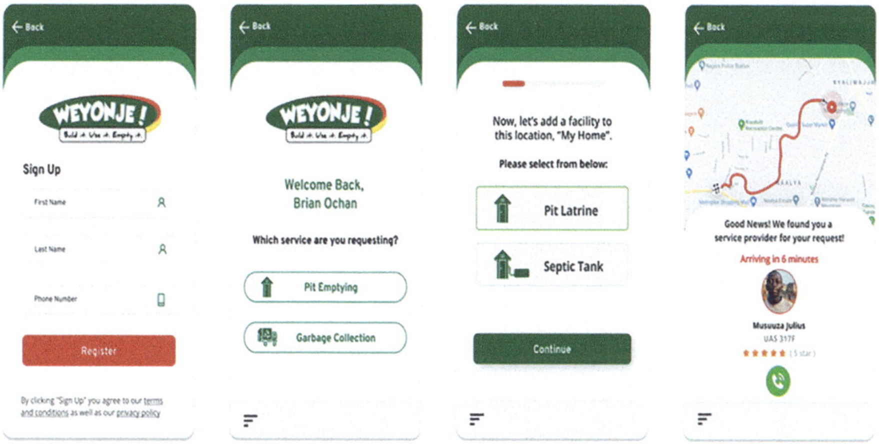

The above images are provided to us by the project owners for their submission.

Chapter 10
Eco-Rehabilitation of the City Park "Chaikovoye Lake" in Kazan, Russia

Abstract In 2022, Kazan embarked on a pioneering initiative to develop a therapeutic park adjacent to a hospital, addressing two key objectives: establishing a rehabilitation area with diverse functional sites and preserving the integrity of the surrounding lake ecosystem. This project not only contributed to the creation of a holistic recovery environment for patients but also fostered an inclusive space for local residents. The therapeutic park, designed with the cooperation of various government authorities and the local community, embodies a unique model of public–private partnership aimed at promoting health and well-being. The park's accessibility has been widely praised by hospital patients, medical professionals, and local residents, providing a multifunctional area for relaxation, physical activities, and social interaction. Moreover, the project has had far-reaching positive impacts, ranging from improved community engagement and enhanced public health to bolstered environmental sustainability and increased local economic activity. The successful integration of nature with healthcare has set a benchmark for future urban development projects focused on human-centered design and ecosystem conservation. The initiative is related to SDGs 3, 6, 8, 9, 11, 14, and 17.

Keywords Therapeutic park · Urban development · Rehabilitation zone · Community cooperation · Healthcare integration · Environmental sustainability · Social impact

10.1 Introduction to Kazan, Russia

One of the most vibrant and historically significant cities in Russia is Kazan, the capital of the Republic of Tatarstan. It is a significant hub for the nation's economy, culture, and science and is situated where the Volga and Kazanka rivers converge. With 1,314,685 residents and a 4% population growth rate, Kazan is still growing as a bustling city that strikes a balance between modernity and tradition. With a population density of 2,232 persons per square kilometer and a total area of 638.40 square kilometers, the city is a vibrant urban area.

A. Cheshmehzangi et al., *People-Centered Innovation*, Urban Sustainability,
https://doi.org/10.1007/978-981-96-5230-3_10

Kazan, one of the most economically important cities in Russia, boasts a diverse industrial base and a GDP per capita of $9,300. The city's main industries are light industry, food production, construction, trade, tourism, machinery, and the chemical and petrochemical sectors. Kazan has become a center of innovation and prosperity as a result of its economic diversity. In comparison to other Russian regions, the city's GINI index of 0.39 indicates a moderate degree of income equality.

Kazan is one of the oldest continuously inhabited cities in Russia, having existed for more than a millennium. It displays a distinctive fusion of Russian and Tatar heritage as a meeting place of European and Asian cultures. Its architectural landmarks, like the UNESCO-listed Kazan Kremlin, which is home to the Annunciation Cathedral and the Kul Sharif Mosque, best illustrate this, signifying the city's religious and cultural cohesion.

Kazan's thriving tourism industry is largely fueled by the city's historic landmarks, vibrant cultural scene, and world-class athletic events. With its hosting of major events like the 2015 World Aquatics Championships, the 2018 FIFA World Cup, and the 2013 Summer Universiade, Kazan has cemented its standing as Russia's "Sports Capital." The city's accessibility and international connectivity are improved by its state-of-the-art infrastructure, which includes the Kazan International Airport and a sophisticated metro system.

In addition to its cultural and economic significance, Kazan is a hub for innovation and education. One of Russia's oldest and most esteemed universities, Kazan Federal University, is located there and promotes scientific and technological development. The city's dedication to advancement is further demonstrated by its investments in sustainability, smart city projects, and urban development. With its robust economy, rich cultural legacy, and progressive development plan, Kazan is a shining example of a Russian city.

10.2 Background and Origins

The history and beginnings of the Kazan Therapeutic Park project can be traced back to a number of environmental, social, and urban planning issues that called for a radical solution. The neighborhood around the hospital was neglected for many years, with little thought given to upkeep or blending it into the larger urban setting. This neglect led to a number of problems, such as the pollution of the nearby lake, which harmed the local ecology and reduced the area's potential as a natural retreat for residents, hospital employees, and patients. Movement within the district was inefficient and fragmented due to the neighborhood's lack of intra-block connections. Furthermore, despite being a vital healthcare facility, the hospital lacked sufficient areas for rehabilitation and recreation, depriving patients and staff of a place to unwind, work out, and recuperate. These elements worked together to produce an atmosphere that was detrimental to wellbeing, especially for those receiving medical care and rehabilitation.

Acknowledging these difficulties, the project's goal was to revitalize the area while meeting social and ecological demands. Preserving and restoring the lake ecosystem was a top priority, making environmental sustainability the focal point of the change. In order to preserve the area's natural integrity and create a peaceful recreational area that would benefit both hospital patients and the neighborhood, the plan called for meticulous landscaping. In addition to trying to reverse pollution, this ecological focus aimed to create a calm and healing environment that might aid in patient recovery.

The establishment of a dedicated rehabilitation area that was adapted to patients' requirements—especially those with restricted mobility—was the second main task. To guarantee that the park became an extension of the hospital's healing process, specific functional areas furnished with rehabilitation simulators had to be included. In contrast to traditional urban parks, this area was created with accessibility and therapeutic purposes in mind, enabling patients to participate in supervised exercise in a relaxing and encouraging environment. The project presented a novel approach to patient care that combined environmental wellness with medical support by incorporating rehabilitation infrastructure into a natural setting.

In addition to helping patients, the project aimed to give hospital employees and nearby residents a place to relax and play sports. Because they frequently work in high-stress environments, medical professionals needed a place to relax and exercise to promote their wellbeing. In order to create a well-balanced environment that supported both physical and mental health, sports facilities and relaxation areas were included. In addition, making the area accessible to the public improved ties between the hospital and the neighborhood, encouraging a feeling of belonging and shared responsibility for maintaining the area.

Therefore, the project's roots are in a multifaceted approach to urban renewal, where community involvement, healthcare infrastructure, and ecological restoration came together to create a sustainable and welcoming public area. The Kazan project is a testament to the effectiveness of combining urban planning, nature, and health by tackling environmental issues, enhancing district connectivity, and establishing necessary rehabilitation and recreation facilities.

10.2.1 Key Innovative Aspects

The Kazan Therapeutic Park project distinguished itself as a model for combining healthcare, urban planning, and community involvement by introducing a number of significant innovations. The project's dual-purpose design, which combined therapeutic and recreational purposes in one urban area, was one of its most innovative features. The Kazan initiative was distinctive in its systematic approach to serving hospital patients and the general public, even though parks close to medical facilities are not unusual. The park's accessibility and openness promoted inclusivity by allowing rehabilitation activities to happen in a communal, natural setting as opposed to being restricted to a medical facility. In addition to helping patients recover in a vibrant, encouraging environment, this strategy improved relationships between the hospital and the neighborhood by promoting communication and assistance amongst members of the community.

The collaborative model of development, which required close collaboration between several stakeholders, was another noteworthy innovation. This initiative brought together representatives from the healthcare industry, local government, academic institutions, and neighborhood residents, in contrast to traditional urban development projects that are frequently led exclusively by municipal authorities. Interdepartmental meetings were conducted on-site during the planning and implementation stages, enabling ongoing input and modifications to guarantee that the park satisfied the needs of all visitors.

The design and operation of the park were actively influenced by medical experts, including specialists from the Kazan State Medical Academy's Department of Rehabilitation and Sports Medicine. In order to guarantee that the park offered patients with different mobility levels real therapeutic benefits, their knowledge directed the choice of rehabilitation equipment, the layout of the exercise areas, and the overall spatial arrangement. The project set a standard for upcoming health-centered urban developments by involving patients, medical professionals, and urban planners in a collaborative decision-making process.

Another significant innovation that set the park apart from traditional public areas was its functional zoning. This project included separate areas for physical therapy and rehabilitation, in contrast to typical urban parks that emphasize leisure and aesthetics. Exercise stations with simulators made for people with limited mobility were among the specialized spaces made for patients in need of structured physical activity. By placing these facilities in a green, outdoor environment, patients were able to participate in rehabilitation exercises and take advantage of the psychological and physical benefits of nature, which helped to promote a more comprehensive healing process. This strategy supported the notion that urban landscapes can be planned to aid in medical treatment and recovery, which is consistent with recent studies on the beneficial effects of green spaces on health outcomes.

The project's long-term sustainability plan was another innovative feature. After it was finished, the park was given to the "Directorate of Parks and Squares," which is in charge of managing the upkeep and continuous improvement of Kazan's public green areas. This made sure the park would receive ongoing maintenance and wouldn't eventually become neglected. In order to keep the park lively and interesting for all visitors, the Directorate also actively plans cultural, athletic, and educational events there. A change in urban planning from one-time infrastructure investments to continuous, community-centered programming that improves the physical and social environment is reflected in this management model. Kazan showed their dedication to long-term urban sustainability and the ongoing enhancement of community well-being by incorporating the park into the city's larger network of public areas.

Its influence on urban governance and social cohesion was arguably the project's most revolutionary feature.

The park's establishment represented a new paradigm of public-authority collaboration in which residents actively shaped their urban surroundings. Residents became more involved in the park's upkeep and success as a result of this participatory approach, which promoted a sense of shared responsibility and ownership. Furthermore, the partnership between city planners and healthcare experts demonstrated the increasing acceptance of interdisciplinary approaches to urban planning. The project demonstrated how cities could become more inclusive, flexible, and responsive to the various needs of their citizens by dismantling the conventional divisions between community development, urban planning, and health.

One of the first examples of how urban areas can be redesigned to support social and medical purposes is the Kazan therapeutic park. The project established a new benchmark for health-focused urban design by incorporating rehabilitation infrastructure into a public park, incorporating a broad spectrum of stakeholders in its creation, and guaranteeing long-term sustainability through institutional support. Its success highlights how cities can use green spaces as vital elements of public health and well-being in addition to being recreational areas. In addition to improving the quality of life for patients and residents, the project's creative approach has shown how well-considered urban interventions can have long-lasting positive social and environmental effects.

10.2.2 Desired Outcome and Measurement

10.2.2.1 Desired Outcome

The project was designed with a multifaceted vision, aiming to provide a comprehensive rehabilitation environment for hospital patients, a stress-relief space for medical professionals, and a recreational hub for local residents. The project aimed to establish a healing environment that would promote both physical and mental well-being by combining recreational and therapeutic activities into one urban setting. By giving patients access to green spaces, fresh air, and recovery-focused exercise areas, the park provided an alternative to traditional indoor rehabilitation. The area provided medical professionals with a haven where they could relax and lessen the stress of their jobs, which in turn enhanced their performance and general well-being. The park gave locals a chance to participate in sports, outdoor recreation, and social interactions, which improved their sense of community and promoted a healthier urban way of life.

The project addressed long-standing environmental concerns by restoring and conserving the lake ecosystem in addition to its benefits for humans. The lake had experienced ecological degradation, weed growth, and progressive shallowing over time, endangering biodiversity and reducing the area's natural beauty. The project made sure that ecological and human health were given top priority by integrating environmental restoration into the larger rehabilitation program. In order to support the lake's long-term sustainability, efforts were made to reduce sedimentation, enhance water quality, and encourage the growth of native plant species. In addition to improving the park's aesthetic value, these interventions strengthened the urban ecosystem's resilience and balance.

The project was carried out in phases, and by August 2021, some landscaping had been finished, making parts of the park open to the public. After all planned interventions were completed, full operational capacity was attained in the summer of 2022. Since then, the park has flourished as a place for rest, recuperation, and recreation, serving as a model for combining community involvement, urban sustainability, and healthcare. The initiative's success shows that similar strategies could be used in future projects, showing how urban areas can be planned to benefit both people and the environment.

10.2.2.2 Measurement

The effectiveness of the therapeutic park project is assessed through a comprehensive measurement framework, capturing its impact on hospital patients, medical staff, and local residents. By integrating clinical, psychological, and observational data, the project evaluates both health outcomes and overall well-being improvements across different groups (Fig. 10.1).

For hospital patients, the park serves as an extension of the rehabilitation process, particularly benefiting those undergoing treatment for diseases of the central nervous

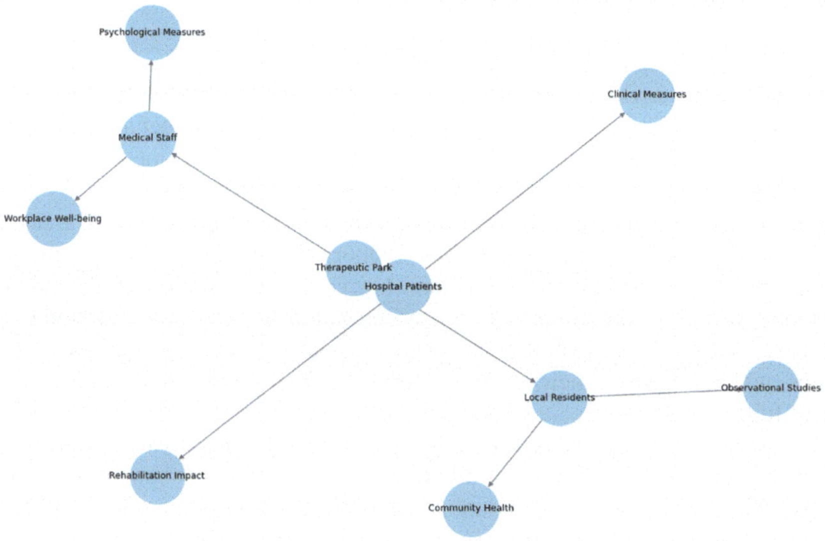

Fig. 10.1 Summary of measurement and key aspects of the project, including identifying the main actors and key stakeholders involved in the project. This also visually represents the measurement framework by showing the relationships between the therapeutic park, different beneficiary groups, and the key evaluation methods used to assess impact

system, musculoskeletal system, and peripheral nervous system, as well as individuals recovering from somatic illnesses. A variety of clinical and instrumental measures, such as the Rivermead Mobility Index, Barthel Index, Functional Independence Measure (FIM), Visual Function Measure (VFM), and SHRM scale, are used to gauge how effective rehabilitation is. Furthermore, Rankin Scale evaluations, six-minute walking tests, and sophisticated medical imaging methods like MRI, EEG, and randomized controlled trials (RCTs) are all used in condition monitoring. By comparing patient progress before and after using the park's therapeutic infrastructure, these assessments enable experts from the multidisciplinary rehabilitation team to provide measurable proof of improvements in mobility, stamina, and general physical health.

The park is essential for medical staff, especially the more than 1,200 workers (including more than 300 doctors), in reducing work-related stress and avoiding emotional burnout. The SAN (Subjective Well-being, Activity, and Mood Scale), Spielberger-Khanin Anxiety Scale, and Beck Depression Inventory are among the well-known psychological tests and scales that hospital psychologists use to assess their patients' wellbeing. These evaluations show how well the park promotes relaxation, lowers stress levels, and improves work-life balance for medical professionals by tracking changes in emotional states over time. The park has emerged as a crucial element in enhancing healthcare workers' job satisfaction and mental toughness by offering a green and healing setting.

The park has given locals access to a useful public area that fosters social interaction, physical exercise, and overall community well-being. Observational studies and surveys show a significant increase in outdoor activity participation and a higher frequency of park visits for social gatherings, sports, and recreation, despite the lack of formalized measurement tools for this group. The project's role in promoting preventive healthcare at the community level has been strengthened by the inclusion of sports facilities, walking paths, and relaxation areas, which have all encouraged healthier lifestyle choices. The project offers a data-driven approach to assessing its impact by utilizing both qualitative and quantitative measurement tools, guaranteeing that its success is not only apparent but also quantifiable. The success of incorporating natural areas into urban healthcare and community well-being strategies is demonstrated by this evidence-based approach, which can be used as a model for rehabilitation parks in the future.

10.2.3 Implementation

10.2.3.1 Principal Parties and Partners

The successful implementation of this project was made possible through the coordinated efforts of multiple principal parties and partners, each contributing their expertise and resources to ensure its realization. The initiators of the project included the Executive Committee of Kazan, the Foundation Institute of Urban Development of the Republic of Tatarstan, and the Ministry of Construction, Architecture, and Housing and Communal Services of the Republic of Tatarstan. Additionally, the Ministry of Health, the Executive Committee of Kazan, and City Clinical Hospital No. 7 played a central role in integrating healthcare and rehabilitation priorities into the project's development.

Important healthcare stakeholders, such as patients, hospital employees, and representatives of the Kazan State Medical Academy's Department of Rehabilitation and Sports Medicine, offered insightful opinions on how the project could best meet community needs. Their input was crucial in determining the project's layout and operation, guaranteeing inclusivity, accessibility, and the advancement of health and wellbeing. An essential part of the project's execution was public engagement. By visiting the architects' office, calling a dedicated phone line, or posting questions, comments, and concerns on the city's website, citizens were urged to actively engage in the process. This candid discussion promoted openness and allowed for ongoing public monitoring of important areas, like park enhancements and waterfowl habitat preservation. The project's direction was improved by the residents' active participation, which made sure that it matched community expectations as well as municipal objectives.

Furthermore, cooperation with landscape architects, urban planners, and environmental experts made sure that ecological preservation and sustainability were incorporated into each phase of execution. Their knowledge was especially important when it came to creating green areas that supported biodiversity while also improving the urban environment. The project demonstrated an inclusive urban development model that balanced community well-being, environmental responsibility, and infrastructure improvements by forging solid partnerships and taking public input into account.

10.2.3.2 Resources for Implementation

The development and implementation of the project were carried out within the framework of the federal initiative *Formation of a Comfortable Urban Environment*. Funding for the project came primarily from the budget of the Republic of Tatarstan, with a total investment of $2.5 million over two years. These financial resources were allocated to various components, including infrastructure development, environmental restoration, accessibility improvements, and public engagement activities.

As the federal project's principal curators, the Ministry of Construction and the Foundation Institute of Urban Development of the Republic of Tatarstan made sure that the project's design and implementation complemented national urban development plans. Through the External Improvement Committee, the Executive Committee of Kazan was instrumental in overseeing the restoration of Chaikovoye Lake Park. This required choosing contractors, managing construction operations, and making sure that sustainability guidelines and project deadlines were followed.

To guarantee that the park's restoration complied with medical and rehabilitation requirements, the Ministry of Health and City Clinical Hospital No. 7 of Kazan also made contributions to the project's healthcare-focused elements. This included creating accessible recreation areas, barrier-free walkways, and wellness-oriented infrastructure to benefit patients, medical staff, and the public at large. A phased strategy was used to maximize resource use. Site preparation, environmental assessment, and obtaining the required permits were the main priorities of the first phase. Improvements to the lake's water quality, lighting, and pathway construction were all part of the second phase's infrastructure. The last stage placed a strong emphasis on community engagement programs, landscaping, and accessibility improvements.

Implementation also benefited greatly from in-kind contributions. Local companies and volunteers helped with the greening initiatives, while architectural firms, landscape designers, and environmental specialists contributed their knowledge of sustainable urban design. For project monitoring, cutting-edge digital tools were also used, enabling real-time evaluation of environmental impact and construction progress. Long-term community benefits were fostered by the project's successful transformation of Chaikovoye Lake Park into a more accessible and environmentally sustainable public space through the effective integration of financial, institutional, and human resources.

10.2.3.3 Tools and Methods Applied

The implementation of the project followed a systematic and multidisciplinary approach, integrating ecological restoration, accessibility enhancements, and specialized rehabilitation methodologies. In order to ensure a healthier ecosystem for the environment and park visitors, the first phase of implementation concentrated on the ecorehabilitation of Chaikovoye Lake. Construction debris was cleared from the lakebed and silt bottom sediments were carefully dredged in order to restore ecosystem balance and water quality. One important strategy to improve the lake's self-purification process and restore it to a more natural state was the introduction of macrophytes, which are self-sustaining aquatic plants. This phase was overseen by environmental specialists who used scientific monitoring tools to evaluate improvements in biodiversity and water quality over time.

The establishment of seven unique therapy zones, each intended to support a different facet of mental and physical rehabilitation, was a fundamental element of the project. To guarantee accessibility for individuals with disabilities and limited mobility, specialized rehabilitation equipment was placed throughout these zones. Wheelchair-accessible walkways, sensory gardens, and rehabilitation-specific exercise stations were all part of this infrastructure. The "7 Sources of Health" concept served as the foundation for the entire rehabilitation project and informed the park's functional design. Evidence-based therapeutic techniques were used to create each of the seven zones, including:

- Aerotherapy Zone: Open spaces with fresh air circulation, promoting lung health and general well-being.
- Active Physical Rehabilitation Zone: Equipped with exercise installations to support mobility restoration and strength training.
- Occupational Therapy Zone: Hands-on activities aimed at improving fine motor skills and cognitive functions.
- Terratherapy Zone: Direct interaction with nature, encouraging relaxation and mental recovery.
- Aerophytotherapy Zone: Incorporating natural phytochemical benefits from coniferous and medicinal plants.
- Bibliotherapy and Quiet Pier (Auto-Relax Zone): A serene environment for relaxation, meditation, and reading therapy.
- Social Therapy Zone: Designed for public engagement, outdoor cinema screenings, and community-building events.

By carefully positioning seating areas around the lake, the landscape therapy zone was created, enabling guests to fully enjoy the outdoors. In addition, there were designated spaces for creative master classes and art-based rehabilitation sessions, as well as an art space for psammotherapy (sand therapy). Using medicinal tonic herbs like lavender, sage, peppermint, and catnip, an aroma oasis was created for sensory-based rehabilitation. Additionally, this area offered therapeutic respiratory benefits due to its ionized air treatment, which was enhanced by the presence of coniferous trees.

A greenhouse and outdoor gardening space were created in order to combine horticultural therapy with sustainable food therapy. Wheelchair users could grow strawberries and honeysuckle on two-level platforms, and patients and guests could participate in therapeutic gardening by growing plants like onions, parsley, and dill. Chaikovoye Lake Park is now a model for health-centered urban environments thanks to the project's successful implementation of a comprehensive and sustainable rehabilitation space through a combination of ecological restoration, creative therapy zones, and inclusive urban design.

10.3 Obstacles and Solutions

The lake's pollution, which had accumulated over time as a result of residential and industrial waste, was one of the main challenges encountered during the project. There were serious aesthetic and environmental problems as a result of this pollution. Over 4.8 thousand tons of trash were cleared from the lake's bottom as part of a comprehensive eco-rehabilitation effort to address this. Furthermore, macrophytes were planted along the shore. Because they encourage natural filtration and help to rebuild the ecosystem, these aquatic plants are crucial to keeping the lake clean. These plants provided a natural remedy for the pollution issue and were essential to the lake's long-term ecological sustainability.

The poor quality of the soil surrounding the lake, which had been significantly impacted by human activity over the years, presented another major obstacle during the rehabilitation process. Due to the soil's inability to sustain the growth of plants and trees, numerous previous attempts to green the lake had failed. In order to get around this, the project team chose to import a lot of fertile land and improve the soil with silt from the lake's bottom. This method helped restore the area's ecological balance in addition to offering a nutrient-rich planting base. The project was successful in bringing in a wide range of unusual trees and shrubs, which have thrived and added to the park's green areas.

Another challenge in the project's early phases was involving locals. At first, many locals were ignorant of the project's specifics and possible effects on the neighborhood. The project team overcame this obstacle by putting in place a thorough communication plan that included disseminating information via social media, printed and electronic news sources, the city's official website, and neighborhood pamphlets. These initiatives promoted increased community involvement in the project's development and raised public awareness. The project team was able to foster public interest and support by keeping residents informed and involved, which resulted in increased involvement in the planning and feedback process.

The rehabilitation project involved numerous structures, which presented an additional challenge. Since numerous departments and organizations were in charge of various facets of the project, it was crucial to make sure that coordination and communication ran smoothly. A specialized project team comprising members from every pertinent department was formed to address this. This team collaborated, maintained continuous communication, and made sure that every stage of the project was carried out as planned. Clear lines of communication and strong interdepartmental cooperation allowed for quick problem solving and kept the project on schedule the entire time.

Despite these significant challenges, the project successfully overcame obstacles through a combination of innovative environmental solutions, effective public outreach, and strong inter-departmental collaboration. The result is a transformed park that not only improves the ecological health of Chaikovoye Lake but also creates a more inclusive and sustainable public space for the community.

10.3.1 Achieved Outcome

Both patients and the larger community have benefited greatly from the rehabilitation project since it began two years ago. The customized rehabilitation programs provided at the health and wellness park "7 Sources of Health" have helped more than 6,500 patients with limited motor capabilities, especially those recuperating from illnesses and injuries like musculoskeletal system disorders (fractures, spinal injuries, endoprosthetics), neurological conditions (stroke, traumatic brain injuries), and internal organ diseases (pneumonia, heart attacks). To improve their physical, mental, and emotional well-being, these patients have participated in extensive, customized therapy.

This rehabilitation program has a remarkable, quantifiable impact. Significant improvements in the patients' conditions are shown by clinical and instrumental indicators. For instance, patients' independence in daily activities increased by 32.4%, and their psycho-emotional state improved by 28.6%, indicating improved mental health outcomes. Specifically, self-care skills increased by 21.8%, which decreased the need for outside help. Additionally, the project improved the quality of life for 68.3% of participants by reducing the primary disability rate by 2.4–4.7%. These results highlight how well therapeutic interventions can be incorporated into a park or other natural, encouraging setting.

The workforce in the community has also benefited from the project. The initiative's wider health benefits in preventing long-term disability and promoting recovery were highlighted by a 10.4% decrease in newly recognized disabilities among working-age individuals when compared to 2017. Socially, the park has developed into a gathering place for the hospital's disabled patients. More than 200 patients enjoy spending time outside each day in the seven functional zones of the park, which helps them recover more quickly and allows them to be discharged two to three days earlier than they would have otherwise. The park's therapeutic atmosphere and the all-encompassing strategy that blends physical rehabilitation with nature-based therapy are responsible for this quicker recovery.

Additionally, the project has greatly improved the health of the medical staff. Medical personnel can relax after long shifts when they have access to an accessible recreation area. Their mental and physical health have improved as a result of playing sports like football, basketball (with two ring levels), and other training exercises; their mental and physical health have increased by 18.9% and 34.7%, respectively. Better employee retention, reduced burnout rates, and a more encouraging work atmosphere have resulted from this.

In addition to providing a tranquil haven and a range of community events arranged by the Directorate of Parks and Squares of Kazan, the park has developed into a welcoming and easily accessible recreational area for the locals, all within walking distance. In addition to creating a much-needed green area for socializing and recreation, this has promoted increased community cohesion. All things considered, the project has had a major social impact, improving healthcare outcomes, raising the standard of living for both patients and locals, and offering other cities a model of inclusive urban design to follow.

10.4 Lessons Learned

Future urban health initiatives can benefit from the lessons learned from the Chaikovoye Lake Park rehabilitation project. During the COVID-19 pandemic, one of the most important realizations occurred when the park played an unexpected but vital role. The facility next to the park was converted into a makeshift hospital for

COVID-19 patients. Despite the difficulties caused by the pandemic, the well-kept park setting was a great help in encouraging patients' and healthcare professionals' recuperation. Spending time outside in the park's functional zones allowed patients to experience the therapeutic benefits of nature, hastening their recovery. Simultaneously, healthcare professionals were able to use the park's recreational spaces to recharge after demanding shifts, significantly improving their mental and physical well-being. This experience highlighted the healing power of green spaces in healthcare settings, reinforcing the need for integrating nature into urban healthcare infrastructure.

The project's reproducibility was another important lesson. The effectiveness of creating green spaces and landscaping in medical facilities showed that comparable projects could be carried out in other Russian cities and abroad. Incorporating accessible outdoor areas and therapeutic parks into medical settings, whether in urban or suburban settings, can significantly improve staff and patient health outcomes. Given the growing emphasis on mental health and well-being in urban planning worldwide, there is enormous potential for scalable replication.

The project's focus on inclusive participation was equally significant. Involving a wide range of stakeholders, from patients and local residents to urban planners and healthcare professionals, turned out to be both beneficial and necessary. The varied contributions of the public, community members, and experts were crucial to the project's success. In addition to fostering community support, involving locals in the planning process made sure the design reflected the needs of those who would use it most. A sustainable and adaptable model that could be adjusted for different environments was produced as a result of this cooperative approach, which improved interdisciplinary cooperation.

A roadmap for upcoming initiatives aiming at improving public health and well-being through careful urban design has been provided by the comprehensive approach that integrated environmental restoration, healthcare services, and community engagement. The initiative's benefits—better patient recovery rates, enhanced physical and mental well-being for medical personnel, and a more involved community—emphasize the potential of incorporating nature and health into urban planning. In conclusion, the Chaikovoye Lake Park project has not only produced immediate health benefits but has also established a model that can be replicated in the future, highlighting the significance of thoughtfully planned green spaces in supporting both mental and physical recovery in urban settings.

Related SDGs: Goals 3, 6, 8, 9, 11, 14, and 17.

Acknowledgements We explicitly thank the project owners based in Kazan, Russia, for their participation and the provision of materials for the 6th Guangzhou International Award for Urban Innovation. We also congratulate them on their achievements in the 2023 cycle of awards.

The above images are provided to us by the project owners for their submission.

Chapter 11
Local SDG Implementation: A Management Tool and the Large Area-Transformation into a SDG-Park in Mannheim, Germany

Abstract The City of Mannheim's management framework incorporates the 17 Sustainable Development Goals (SDGs) in a significant way, demonstrating the city's dedication to social and environmental advancement. Mannheim began implementing a comprehensive plan in 2015 that promoted sustainable development and emphasized the significance of striking a balance between social, environmental, and economic advancement. In order to localize the 17 UN Sustainable Development Goals, Mayor Kurz launched a thorough citizen participation process in 2018. This process resulted in the creation of the Mannheim 2030 strategy, which was developed in conjunction with locals. The effective alignment of local initiatives with global efforts to achieve the SDGs, as well as the promotion of international comparability and knowledge exchange, remain significant challenges. In response, Mannheim unveiled the 2022 annual report (VLR), establishing the VLR as a management tool that is crucial for guiding local policy and actions in addition to being a reporting tool. In order to further ensure accountability, transparency, and a comprehensive approach to sustainable development, this report—which will be released annually—complements the city's mission statement, target systems, municipal budget plans, and a list of ongoing projects. Mannheim hopes to make a significant contribution to the global SDG agenda while customizing solutions for its distinct urban setting by fusing local action with global frameworks. In Part 1, the initiative is related to all the SDGs, and in Part 2, the initiative/project is related to SDGs 3, 11, and 13.

Keywords Sustainable development · SDGs · Mannheim 2030 · Citizen participation · Local action · Global agenda · Voluntary local review (VLR)

11.1 Introduction to Mannheim, Germany

The southwestern German city of Mannheim is making great progress in incorporating the Sustainable Development Goals (SDGs) into its local government. Mannheim is one of the major cities in the state of Baden-Württemberg, with a population of 329,608 and a noteworthy 7.4% growth rate. It is renowned for its industrial

A. Cheshmehzangi et al., *People-Centered Innovation*, Urban Sustainability,
https://doi.org/10.1007/978-981-96-5230-3_11

prowess and dedication to sustainability. Launched in 2022, the city's annual report on SDG achievement represents a significant innovation in local governance: turning the Voluntary Local Review (VLR) into a potent management tool that facilitates decision-making and advances sustainable development, rather than merely a monitoring tool. In addition to being a reporting procedure, Mannheim's SDG strategy is a component of a larger initiative to integrate sustainability into every facet of urban life.

With 2,274 inhabitants per square kilometer, Mannheim has a surface area of 144.96 square kilometers. This relatively high density underscores the city's role as a hub of economic and social activity in the region. Mannheim's strong industrial base is a major contributor to its thriving economy, which has a GDP per capita of $3,230.43. The city is well known for being at the forefront of industries like manufacturing, technology, and energy, especially in the chemical, automotive, and industrial machinery sectors. These industries are essential to the city's prosperity and serve as the cornerstone of its efforts to develop into a sustainable urban hub.

Since the UN adopted the SDGs in 2015, they have been firmly incorporated into Mannheim's strategic framework. Under Mayor Kurz's direction, the city began a comprehensive citizen participation process in 2018 with the goal of localizing the 17 SDGs. Through this initiative, the public was able to directly influence the Mannheim 2030 strategy, which is a detailed plan for the city's growth over the ensuing ten years. With the public and stakeholders actively involved at every stage, the 2030 strategy aims to create a more resilient, inclusive, and sustainable future.

A significant turning point came in 2022 when Mannheim introduced its annual SDG report as part of its Voluntary Local Review (VLR). Unlike traditional reports that merely monitor progress, the VLR in Mannheim serves as an active management tool. This new approach allows the city to not only track its achievements and challenges in real-time but also adjust its policies and strategies accordingly to ensure that the SDGs remain central to all decision-making processes. The VLR, which will be published annually, complements the city's mission statement, target systems, and municipal budget plans. This framework is designed to provide a holistic view of how Mannheim is working towards sustainable development goals while maintaining transparency, accountability, and public engagement.

Understanding Mannheim's industrial base is essential to comprehending its SDG strategy. The city, one of Germany's industrial superpowers, is well aware of the need to switch to more sustainable production practices and greener technologies. The city's strategy for implementing the SDGs is centered on integrating energy-efficient technologies, green industry practices, and the advancement of circular economy concepts. Additionally, Mannheim is dedicated to promoting social inclusion by tackling income inequality (as indicated by its GINI index of 0.3) through policies that support equitable access to economic opportunities, healthcare, education, and housing.

The development of green spaces, public transportation as a way to cut carbon emissions, and other urban renewal projects all demonstrate the city's strong emphasis on environmental sustainability. Mannheim has also concentrated on making sure that all of its urban development initiatives are in line with the SDGs' climate action targets in order to create a city that is both environmentally conscious and economically prosperous.

The government of Mannheim collaborates closely with a wide range of stakeholders, such as nearby companies, academic institutions, and civil society organizations, to further propel the SDG initiative's success. By ensuring that the city's SDG initiatives are in line with the requirements of all societal segments, these partnerships promote an urban environment that is more inclusive and equitable. By holding the annual VLR, Mannheim has established itself as a role model for other German and international cities, showing how local governments can be crucial to accomplishing the SDGs while tackling the particular opportunities and challenges of their urban setting. The incorporation of the SDGs into routine governance guarantees that sustainability will continue to be a fundamental tenet driving Mannheim's growth in the years to come as the city continues its transition to a sustainable future.

11.2 Background and Origins

Mannheim's efforts to integrate the Sustainable Development Goals (SDGs) into its local governance have advanced significantly with the establishment of the annual SDG report and, consequently, the Voluntary Local Review (VLR). The VLR is a crucial instrument for the open, inclusive, and responsible administration of Mannheim's sustainable development plan; it is not just a way to monitor advancement. In order to make sure that every facet of city life, from social services to infrastructure, is in line with the overall objectives of the SDGs, the city collaborates with all pertinent municipal departments and businesses to compile the report annually.

The annual report, once finished, is given to the municipal council and included in the formal agenda for public deliberation and decision-making. This approach guarantees that public policy will continue to be centered on the Mannheim 2030 vision, which provides a thorough framework for accomplishing the SDGs locally. The city has pledged to make this report publicly accessible in order to promote increased openness and public participation. Mannheim's commitment to the larger international goals is further demonstrated by the annual review process, which not only shows the city's progress in reaching the SDGs but also highlights its contributions to the global SDG agenda. Furthermore, Mannheim is establishing a standard for other cities hoping to successfully localize the SDGs by presenting the report in this transparent and approachable way.

Additionally, the VLR gives local stakeholders and the general public a chance to interact with the city's plans for sustainable development, provide input, and promote discussion about the opportunities and difficulties associated with implementing the SDGs. Mannheim's dedication to establishing a more inclusive, egalitarian, and ecologically conscious urban environment is strengthened by this participatory approach, which guarantees that the community's voice is represented in the city's advancement toward sustainability.

The growing understanding that crises like the COVID-19 pandemic, inequality, and climate change were interconnected problems requiring long-term, coordinated solutions is where Mannheim's emphasis on SDG integration got its start. The persistent crises were particularly highlighting the shortcomings of conventional governance models, which were frequently reactive and lacked the clarity and vision required to handle the intricate problems that cities face. Like many other cities worldwide, Mannheim struggled to strike a balance between immediate fixes and long-term sustainability objectives. The current approach to sustainable development was reevaluated after it was realized that the city was not on track to meet the 17 SDGs by 2030. Mannheim recognized the need for a more focused, organized strategy in light of the world's growing instability, one that would allow it to both adjust to the quickly shifting environment and make quantifiable progress toward the SDGs. As a result of this insight, a more structured framework for SDG localization was developed, with the yearly VLR report acting as a key instrument for monitoring progress and modifying plans as necessary.

The initiative's fundamental goal is to transcend the constraints of conventional crisis management and shift toward a long-term, proactive, and flexible governance model. This new strategy, which stresses specific objectives, doable tactics, and efficient governance models to stay on course both locally and globally, is reflected in the VLR report. As the need for a clear, uncomplicated framework to link local initiatives with global objectives grew, Mannheim's leadership recognized the annual report as the key to achieving this.

Fundamentally, the VLR makes sure that the city's efforts to achieve the SDGs are incorporated into a continuous, sustainable management process rather than being fragmented or reactive. This change represents a turning point for the city as it aims to go beyond simply reacting to problems to methodically pursuing a more resilient, sustainable future. Mannheim is establishing the groundwork for a governance model that can successfully navigate upcoming obstacles and make a significant contribution to the global SDG agenda while simultaneously adjusting to the needs of the local community by focusing its efforts on specific, quantifiable goals.

11.2.1 Key Innovative Aspects

The establishment of an Annual SDG Report, which functions as both an active tool for strategic governance and a monitoring tool, is among Mannheim's most inventive approaches to the Sustainable Development Goals (SDGs). This report provides an overview of the city's efforts and progress toward its sustainability goals by compiling all city-related activities that are in line with the SDGs and the Mannheim mission statement. The city can continuously monitor its progress and make necessary strategy adjustments to stay on track to meet the global targets set for 2030 thanks to the annual compilation of activities and accomplishments.

Mannheim's collaborative approach is a particularly novel aspect of this process. The city's SDG strategy, in contrast to conventional top-down management systems, promotes active engagement from a variety of stakeholders, including independent organizations, societies, all city departments, and other community partners. This cross-sector cooperation guarantees that every department directly contributes to the city's overall sustainability agenda and cultivates a shared commitment to accomplishing the SDGs.

This cooperative process is largely coordinated by the Democracy and Strategy Department, which is in charge of integrating the SDGs into all facets of the city's governance. To promote a transparent exchange of ideas and guarantee alignment between individual goals, indicators, and strategies, this department holds yearly discussions with all departments and affiliated organizations. The city avoids fragmentation and establishes a cohesive, citywide approach to sustainability by doing this. By ensuring that every department's efforts support the larger SDG framework, this integrated model strengthens a shared future vision.

The emphasis on common objectives and quantifiable metrics is another significant innovation. Mannheim's method places more emphasis on quantitative indicators than qualitative reports, which makes it possible to track progress objectively. This guarantees that the process of implementing the SDGs is accountable, transparent, and data-driven. It is easier to identify areas that require improvement or additional investment when departments are required to set clear, quantifiable goals and periodically assess their progress.

This methodology's ability to be flexible in responding to new challenges makes it especially effective. The city's experience has demonstrated that, as local and global conditions change, concentrating on long-term sustainability necessitates constant adaptation. The dynamic nature of Mannheim's approach allows it to react swiftly to unanticipated crises like public health issues or climate events, guaranteeing that SDG initiatives stay resilient and relevant in the face of uncertainty.

Additionally, innovation is promoted at all governmental levels by Mannheim's collaborative model. The city can use the knowledge and perspectives of several industries, such as healthcare, education, urban planning, and environmental sustainability, to create more innovative answers to challenging urban problems with the help of varied stakeholders. This innovation involves rethinking governance itself as well as coming up with fresh approaches to achieving the SDGs. Because of Mannheim's dedication to transparency, inclusivity, and stakeholder engagement, a shared responsibility culture has been established, making sustainability a group endeavor driven by the community. Ultimately, Mannheim's creative strategy shows that localizing the SDGs involves more than just putting global goals into practice locally; it also entails rethinking governance frameworks to promote inclusive, sustainable, and long-term growth. Mannheim has set an example for other cities to follow with its integrated approach and annual report, demonstrating the importance of cooperation, data transparency, and flexibility in reaching the SDGs in a significant and lasting manner.

11.2.2 Desired Outcome and Measurement

11.2.2.1 Desired Outcome

The initiative's main objective is to make the Voluntary Local Review (VLR) and the Annual SDG Report official instruments of governance in the City of Mannheim. This report is intended to serve as a crucial tool for the city's yearly decision-making process beginning in 2023. In order to ensure that the city's actions continuously support the Sustainable Development Goals (SDGs) at the local and global levels, the city hopes to improve transparency, accountability, and strategic alignment across all departments by regularly including the VLR on the municipal council's agenda. The city is successfully converting international commitments into tangible, quantifiable actions within the local context thanks to this integration, which helps close the gap between the global SDG targets and local needs.

Showing that local priorities and global goals must coincide and complement one another is one of the initiative's intended results. Mannheim's approach is based on the idea that sustainable development involves more than just fulfilling international commitments; it also entails integrating the SDGs into locals' everyday lives. Mannheim aims to guarantee that local policies, planning, and development initiatives are completely in line with the global agenda by incorporating the SDGs into the city's governance framework. The city's development framework is based on this alignment, which is not just symbolic but guarantees that local and global objectives "speak the same language".

Furthermore, a unified and inclusive system where all sectors—government, corporations, civil society, and citizens—actively participate in accomplishing these objectives is the intended result. Mannheim hopes to normalize sustainability practices and make them an integral part of the city's operations by incorporating SDG-focused governance into the decision-making process. A dynamic, continuous commitment to sustainable urban development will be fostered by the city's ability to evaluate and modify its strategies annually thanks to the VLR, a regular governance tool. In addition to serving as a monitoring exercise, the report is meant to be a tool for continuous improvement, where feedback loops aid in the improvement of strategies to address new issues like social inequality, climate change, and economic resilience.

In order to make sustainable development both an ideal and a reality, the ultimate goal is to position Mannheim as a preeminent example of how cities can successfully localize the SDGs. By incorporating the VLR into the governance framework, Mannheim hopes to encourage other cities to follow suit and demonstrate that it is both essential and feasible to match local governance with international sustainability objectives.

11.2.2.2 Measurement

The Annual SDG Report, which offers a thorough assessment of the city's progress toward accomplishing the SDGs and its strategic objectives, is prepared primarily through the collection and analysis of both quantitative and qualitative indicators. These metrics are crucial for monitoring policy efficacy and making sure Mannheim's initiatives are in line with its long-term sustainability goals. A comprehensive understanding of the city's development and its role in the global SDG agenda is made possible by the indicators gathered, which cover a variety of aspects of urban life.

The Annual SDG Report includes 50 key indicators at the highest level, which give a general picture of the city's performance. These metrics are based on local development priorities as well as international sustainability standards. Among these, standard metrics that provide quantifiable and unambiguous information on environmental, economic, and social factors include CO_2 emissions, the unemployment rate, transmission rates to higher education, and mobility trends. These metrics give Mannheim accurate information on how it is doing in relation to important global goals like SDG 8 (Decent Work and Economic Growth) and SDG 13 (Climate Action).

Mannheim's method is not restricted to conventional, well-known indicators, though. In order to capture the social dynamics and citizen engagement that are essential to sustainable urban development, the city also uses local qualitative indicators. For example, the report includes "satisfaction with citizen participation" and "trust in city government." These metrics are essential for evaluating the success of democratic procedures and participatory governance as well as for comprehending the city's relationship with its citizens. Mannheim makes sure that the human aspect of sustainability—how people perceive and engage with their surroundings—is also closely observed by incorporating such measures.

Mannheim's approach to measuring the SDGs is very thorough since it integrates both quantitative and qualitative indicators, enabling the integration of objective data and subjective opinions. This combination gives a comprehensive picture of the city's development and sheds light on areas that might need improvement, particularly with regard to social cohesion, public trust, and citizen satisfaction.

Additionally, the city's Budget Plan and Mission Statement are closely related to these indicators, guaranteeing that they are not only a reflection of the city's priorities but also integrated into its financial and strategic planning procedures. This integration guarantees that the pursuit of SDGs is consistently supported by city policy and funding, which is crucial for the initiative's long-term viability. More open decision-making is also made possible by the budget's alignment, and the municipal council actively participates in yearly discussions and assessments of these indicators' advancement. Ultimately, Mannheim's measurement system is a flexible instrument that aids the city in evaluating its initiatives in real time, allowing for prompt modifications to its plans and initiatives. In order to maintain both global alignment and local relevance in its sustainability efforts, Mannheim continuously monitors and discusses these indicators to make sure it stays on track to meet the SDGs while simultaneously enhancing the well-being of its citizens.

11.2.3 Implementation

11.2.3.1 Principal Parties and Partners

The cooperation of numerous stakeholders, including both domestic and foreign partners, is essential to the successful execution of Mannheim's SDG initiative. A broad and inclusive approach to citizen participation is essential to the Mannheim 2030 mission because it guarantees that the city's strategy is developed with the needs and aspirations of its diverse population in mind. The mission statement was the result of active contributions from more than 2,500 Mannheim residents, companies, institutions, self-help groups, civil society organizations, universities, and associations. The city was able to gather a variety of ideas and proposals through this participatory process, guaranteeing that the finished plan is thorough, inclusive, and consistent with community values.

The project's citizen engagement component is crucial because it guarantees that the local populace takes ownership of and responsibility for the SDG-related objectives, which facilitates grassroots initiative implementation. Stronger social cohesion and a sense of community are also fostered by this cooperative process, which is crucial for accomplishing the SDGs. As a result, the city's sustainability policies are more credible and effective overall when local actors—from businesses to community leaders—are actively involved.

Another crucial element of the project's success is international cooperation. The City of Mannheim understands that global problems call for global answers. In order to promote knowledge sharing, the sharing of best practices, and cooperative problem-solving, the city has thus partnered with numerous international organizations, municipalities, and networks. Mannheim can contribute to the larger global efforts to achieve the SDGs and make sure that its local initiatives are informed by global trends and innovations by utilizing its international connections. This includes taking part in global platforms like the Urban Thinkers Campus (UTC), which promotes communication and cooperation amongst global urban stakeholders.

Mannheim hopes to establish a feedback loop through these global collaborations that will enable it to improve its tactics and more accurately gauge its progress. By taking part in international discussions, the city can monitor its performance, compare its accomplishments to those of other cities, and find new areas for development. The city's focus on collaboration and comparability supports the notion that local initiatives can spur global change when there is shared commitment, learning, and a concerted effort to address the most important issues facing the globe. The city's SDG initiatives are strengthened by this mix of international cooperation and local involvement, which makes them more flexible, inclusive, and accountable. The collaborative approach also highlights the idea that all facets of society, from individuals to international organizations, should collaborate in order to achieve a common vision of a sustainable future, rather than relying exclusively on governmental authorities to pursue the SDGs.

11.2.3.2 Resources for Implementation

The city's Strategic Management team is principally responsible for the execution of Mannheim's SDG initiative, especially the creation of the Annual Report. This group is in charge of organizing the gathering, processing, and display of information about the city's advancement toward the SDGs. Because the project is incorporated into the city's strategic management framework, the personnel costs for this team are directly related to the project's actual costs. The annual report does not necessitate a substantial financial investment beyond the team's operating expenses because it functions as a conceptual control measure rather than a large-scale infrastructure project.

Personnel engaged in data collection, coordination, and report preparation account for the majority of the project's resource costs. There are no significant additional financial burdens related to the project because the Strategic Management team collaborates closely with the Democracy and Strategy Department, and the report's personnel costs are viewed as a normal part of the city's governance operations. The report instead makes use of the city government's current human resources and infrastructure.

The cooperation with different city departments is another important resource in the report's implementation. In order to ensure that data is gathered reliably and consistently, the Democracy and Strategy Department is essential in coordinating activities across the city's many sectors. The indicators' comprehensiveness, which reflects the SDGs' multifaceted nature and the Mannheim 2030 strategy's interconnected goals, is ensured by departmental collaboration. Collecting pertinent data on a variety of subjects, including social cohesion, economic growth, environmental sustainability, and citizen participation, requires this cross-departmental collaboration.

The Annual Report utilizes current local data systems and procedures in addition to internal city resources. The fundamental frameworks for identifying the pertinent indicators and tracking advancement are provided by the city's mission statement and budget plan. Because the report makes use of the data flows and governance mechanisms that are already in place within Mannheim's city administration, using these established structures reduces the need for additional financial resources.

The project's design places a strong emphasis on financial efficiency due to its conceptual nature. Mannheim guarantees that resources are used efficiently by incorporating the report preparation into the city's continuing management structure, obviating the need for significant outside funding or special budgets. Rather, the project focuses on making the most of the city's current data systems, human resources, and infrastructure, which makes it a sustainable and economical endeavor. This strategy demonstrates Mannheim's dedication to reaching its SDG goals while upholding operational effectiveness and financial responsibility. Additionally, the resource structure facilitates scalability and flexibility. Mannheim can modify its strategy as the project develops in response to lessons learned and changing requirements, guaranteeing that the project stays flexible while maintaining low costs. This guarantees long-term success in the localization of the SDGs by enabling the city's SDG monitoring and management procedures to be continuously improved.

11.2.3.3 Tools and Methods Applied

A range of instruments and techniques are used in the execution of Mannheim's SDG strategy, especially in the creation of the Annual Report, to guarantee precise measurement and efficient administration of the SDGs' advancement. The indicator system, which is essential to the process of monitoring and reporting, is one of the main tools utilized. More than 50 important indicators covering the environmental, social, economic, and governance facets of sustainability make up this system. These metrics include quantitative ones like unemployment rates and CO_2 emissions, as well as qualitative ones like public satisfaction with participation procedures and citizen trust in local government.

Mannheim uses an extensive data management system to gather the information required for these indicators. This system is intended to collect data from various sources inside the city's administrative framework, enabling a centralized and integrated method of monitoring the SDGs' advancement. Making sure that all pertinent departments participate in the data collection process is a crucial responsibility of the Democracy and Strategy Department. A comprehensive picture of the city's progress toward sustainability is provided by this interdepartmental cooperation, which guarantees that the indicators accurately represent the range of its operations.

In order to include the public in the process, the city also uses participatory methods. A great deal of citizen participation went into creating the Mannheim 2030 vision, which serves as the foundation for the city's SDG strategy. Throughout the implementation phase, this engagement has persisted, and the Annual Report provides a chance to further engage the public in conversations regarding the city's advancements. In order to facilitate continuous feedback and cooperation, the public can view the report, participate in online discussions, and attend community meetings.

International cooperation and comparability are two more crucial implementation strategies. Mannheim frequently updates other cities around the world on its progress, evaluating how its initiatives stack up against global best practices through knowledge sharing and benchmarking. Through this cooperative learning process, Mannheim can hone its tactics, spot fresh chances for development, and make sure that its strategy stays in line with the larger global SDG framework.

The process will continue to be open and accountable because of the city's dedication to using transparent governance tools. The Annual Report itself is a tool that ensures that all stakeholders, from city dwellers to foreign partners, have access to information about Mannheim's SDG efforts. It also acts as a communication tool in addition to being a monitoring tool. Mannheim guarantees that the entire extent of its SDG implementation is recorded by combining quantitative and qualitative data, offering a strong basis for continuous enhancements and well-informed decision-making.

In conclusion, Mannheim is able to efficiently manage and monitor its progress toward the SDGs thanks to its utilization of a comprehensive and integrated set of tools and techniques, such as its indicator system, participatory engagement, international collaboration, and transparent reporting practices. These techniques are essential to upholding an inclusive and data-driven strategy, guaranteeing that the city's sustainability initiatives are successful and sensitive to the needs of its residents.

11.3 Obstacles and Solutions

The City of Mannheim encountered several challenges while implementing its approach to the Sustainable Development Goals (SDGs) through its Annual Report. These obstacles, while significant, were met with innovative solutions that have helped the city maintain progress toward its SDG objectives.

Converting international agendas into local initiatives was one of the city's biggest problems. The United Nations created the Sustainable Development Goals (SDGs), which are essentially global in scope and aim to tackle problems such as inequality, climate change, and sustainable economic growth. However, at the local level, these overarching objectives can occasionally feel nebulous, making it challenging for municipalities to pinpoint specific actions that will help them achieve their objectives. This made it difficult for Mannheim to turn these international pledges into workable, regional plans. The city had to close the gap between the high-level global goals and the day-to-day issues that its citizens and government agencies faced.

This issue was compounded by siloed thinking within the city's administration. Various departments, while all working toward similar goals, traditionally operated with their own agendas, resources, and reporting systems. A unified approach to the SDGs was difficult to achieve as a result of this departmental fragmentation. The reports from each department were frequently separate from the SDGs Annual Report. It was more difficult to develop a cohesive plan that directed all city efforts toward a single objective because of this lack of integration. Additionally, given the limited resources available, there was occasionally the impression that the goals were at odds. Initiatives aimed at promoting economic expansion, for instance, might seem to conflict with environmental objectives like cutting carbon emissions. When these opposing objectives were not appropriately aligned, tension resulted and real progress was impeded.

All departments participated in the development and execution of the SDG strategy as part of Mannheim's inclusive, cross-departmental approach to addressing these issues. The Democracy and Strategy Department collaborated with other departments to integrate their efforts and align their objectives by hosting yearly discussions and maintaining continuous communication. By encouraging cooperation and dismantling organizational silos, this approach made sure that all municipal initiatives were aimed at the same overall goals. Additionally, the report was created as a management tool rather than merely a monitoring tool, enabling ongoing feedback and modification to guarantee that all objectives were in line.

Another key obstacle was the perceived conflict between global and local goals. Global commitments can often feel distant or abstract, and transferring them into a localized context presents a significant challenge for municipalities, particularly in the face of limited resources. Mannheim had to balance local priorities with the global agenda while modifying the SDGs to fit its particular context. The city concentrated on making the link between the SDGs and local development priorities clear in order to get past this. Through public participation initiatives like the Mannheim 2030 vision, the city brought the SDGs into the lives of its residents. In addition to making sure that the global objectives were understood locally, this process gave the populace a sense of accountability and ownership. Additionally, by providing lucid, actionable indicators that demonstrated observable progress toward accomplishing the goals, the Annual Report contributed to demythologizing the SDGs.

The national charter implementation plans presented another challenge since they were frequently viewed as appellative in nature, which meant that they focused more on making arguments than on offering concrete recommendations or doable actions for local implementation. Without enough assistance or direction from higher governmental levels, local employees who were entrusted with deciphering and carrying out these plans frequently found themselves fighting alone. In Mannheim's case, the city's independence helped to offset the absence of a robust national support system. The city forged its own route to accomplishing these global goals and established its own internal framework for SDG localization. Mannheim was able to successfully negotiate these difficulties because of its dedication to cooperation and involvement, both internally within city departments and externally with residents and foreign partners.

To sum up, Mannheim's strategy for overcoming challenges in the SDGs' implementation emphasizes the significance of cooperation, localization, and integration. The city has been able to overcome the difficulties of fragmented governance, competing objectives, and scarce resources by coordinating departmental goals, involving citizens, and modifying international commitments to the local setting. By implementing these solutions, Mannheim keeps moving closer to reaching its SDG goals and building a more resilient, inclusive, and sustainable city.

11.3.1 Achieved Outcomes

The introduction of the Annual Report as a tool for tracking the Sustainable Development Goals (SDGs) has produced several positive and measurable outcomes for the City of Mannheim. One of the primary successes is the localization of global SDGs through a clear, district-level assessment of various indicators. These metrics, which are linked to Mannheim's Mission Statement and the SDGs, show how the city is doing on a variety of social, economic, and environmental issues. Crucially, all Mannheim residents have access to the data collected in the Annual Report, in addition to policymakers. This transparency fosters accountability and citizen participation by giving residents a chance to participate in local governance and observe firsthand how their city is doing in achieving the SDGs.

The city has been able to pinpoint particular districts and neighborhoods that require focused interventions thanks to the Annual Report. Through the evaluation of metrics like child poverty, educational equity, accessibility, and cleanliness, the city has been able to determine the most effective areas for resource allocation. For instance, regions with greater rates of child poverty have been identified in the report, suggesting a need for more social services and family support. Additionally, it has pinpointed districts where educational disparities still exist, offering important information about where to focus resources and initiatives to lessen these disparities.

Finding the places where accessibility is still a problem is another important result of the Annual Report. For the elderly and people with disabilities in particular, this is crucial. Mannheim can prioritize projects that enhance infrastructure and make the city more welcoming for all citizens by using the report to monitor accessibility issues. For instance, the city can use this information to guide planning and implement the required changes if specific districts are found to have insufficient options for accessible public areas or public transportation. As a result, the report is crucial in forming laws that guarantee inclusivity and equity for all groups.

The Annual Report's thoroughness and transparency, which enable residents to interact with it directly, are among its noteworthy advantages. Greater public awareness of the difficulties the city faces in accomplishing its SDGs has been facilitated by this transparency. Now, citizens can see where their community is at in terms of achieving particular SDG targets and where more work is needed. Furthermore, residents are empowered to actively participate in discussions regarding policy priorities due to the report's accessibility.

Additionally, the Annual Report has offered a way to assess the success of different projects that have been carried out over the course of the previous year. For example, it enables the city to determine whether policies intended to improve public services, expand educational opportunities, or lessen child poverty have had a discernible effect. Mannheim can make sure it is consistently moving closer to the global SDGs and modifying its approach as needed to meet its long-term objectives by monitoring these indicators year after year. To sum up, the Annual Report has been an essential instrument for Mannheim's evidence-based decision-making process. The city has been able to determine where interventions are most needed, whether that be addressing child poverty, promoting educational equity, or improving accessibility for vulnerable populations, by decomposing data at the neighborhood level. The report's transparency and engagement have improved citizen accountability and participation, bringing the city one step closer to realizing its goal of a more equitable, sustainable, and inclusive future.

11.4 Lessons Learned

The Mannheim 2030 mission statement, alongside the Annual Report, has evolved significantly since its inception, providing valuable lessons that can be applied to future urban governance and SDG localization efforts. One of the key lessons learned has been the importance of continuous adaptation and improvement based on real-time feedback from citizens and global experts. For instance, during an Urban Thinkers Campus event, Mannheim engaged with residents and international stakeholders to reflect on the lessons learned from the Covid-19 pandemic. This process revealed the need for more flexible and attractive public spaces, prompting the city to add a new goal focused on the creation of such spaces in every district. This illustrates the value of citizen engagement and how it can shape a city's strategic direction in real-time, making it more responsive to emerging challenges.

The pandemic also highlighted how important it is that public areas be used as places to unwind as well as useful community centers that foster wellbeing and social cohesiveness. Because of this, Mannheim's public space design has changed to emphasize outdoor lounges where people can congregate in safety, mingle, and participate in group activities. A greater awareness of how public spaces can promote community ties and enhance quality of life has led to a renewed emphasis on urban gardening, parks, and the incorporation of social and gender considerations into public space design. The city's strategy now takes into account the social, emotional, and gender-inclusive dimensions of space in addition to its physical characteristics.

Mannheim's early recognition of the potential for city-to-city learning has been another important lesson learned. While acknowledging that cities around the world must work together to achieve the global goals, the city also recognizes its global responsibility to advance the localization of the SDGs. Mannheim is dedicated to encouraging other cities to establish their own Voluntary Local Reviews (VLRs) and has actively participated in knowledge sharing with other municipalities. When coordinated with local initiatives, these VLRs play a crucial role in promoting significant local action and bolstering international sustainability initiatives. Because it promotes a sense of shared responsibility and support among cities worldwide, this shared commitment to cooperation is essential for accomplishing the sustainable localization of the SDGs.

In conclusion, the significance of flexibility, public involvement, and international cooperation are the main takeaways from Mannheim's experience with the Annual Report and the Mannheim 2030 mission. Mannheim has been able to continuously improve its approach to urban sustainability by paying attention to the needs of its citizens and learning from professionals around the world, guaranteeing that its objectives stay applicable and successful. The city's dedication to promoting public areas, group education, and neighborhood action shows that both local initiative and international cooperation are necessary to meet the SDGs.

Related SDGs: All the Goals, SDGs 1–17

1 NO POVERTY
2 ZERO HUNGER
3 GOOD HEALTH AND WELL-BEING
4 QUALITY EDUCATION
5 GENDER EQUALITY
6 CLEAN WATER AND SANITATION
7 AFFORDABLE AND CLEAN ENERGY
8 DECENT WORK AND ECONOMIC GROWTH
9 INDUSTRY, INNOVATION AND INFRASTRUCTURE
10 REDUCED INEQUALITIES
11 SUSTAINABLE CITIES AND COMMUNITIES
12 RESPONSIBLE CONSUMPTION AND PRODUCTION
13 CLIMATE ACTION
14 LIFE BELOW WATER
15 LIFE ON LAND
16 PEACE, JUSTICE AND STRONG INSTITUTIONS
17 PARTNERSHIPS FOR THE GOALS

MEGA EVENT
BUGA

The above images are provided to us by the project owners for their submission.

Added Information about the Second Part of the Project:

The project is developed into two parts, with the second part focused on the transformation of a 80 hectare former military area into a park in the city as a symbol for the localization of the 17 SDGs. The Federal Republic of Germany promotes the greening of cities with the Federal Gardening Show (BUGA) program. With the departure of the U.S. military bases in 2014 in Mannheim, large areas of the city became vacant.

There were quick responses for allocating big logistics enterprises, which meant a large, sealed area, buildings, relatively few jobs, and a high traffic volume. With the adoption of the SDGs, the City of Mannheim initiated a large citizen participation process that included the localization of the SDGs into a new Mannheim 2030 strategy. Every effort was made to develop the area in terms of an overall ecological and social concept for the whole city. This idea was to be implemented by means of the vehicle of the BUGA. In this way, the topics of "urban greening and fresh air supply to cool down the inner city", "local recreation areas", "biodiversity" and "social housing" could be combined. The idea became clear to dedicate this Federal Gardening Show specifically to the SDGs and thus symbolically and above all concretely demonstrate solutions for achieving them. Through Charter management, the City is screening and identifying relevant international charters, coordinating, adopt them with the councilors, and bring the local relevant items of this charters directly in its management target system of the departments. And it works with the NUA and others.

The decision to organize the Federal Gardening Show (BUGA) program was first made as a citizens referendum and then confirmed by the resolution of the Municipal Council. Subsequently, the strategy was developed in connection with the localization of all 17 SDGs and presented to the municipal council in 2017. In 2023, it was reported that the City is off track in its efforts to achieve the 17 SDGs. The current crises and constant new challenges are causing public administrations to stumble from one crisis to the next. It needs concrete actions in terms of SDG localization, which is reinforced concretely with the BUGA. More importantly, Mannheim is the most sustainable BUGA ever. Environmental and climate protection, resource-saving energy generation, and sustainable food security are the focus. In addition, the BUGA is an experimentation and information field, especially for the topics "regional urban food production" and "clean and climate-friendly energy production in cities".

Within a few years, the BUGA site has been transformed from an 80-hectare sealed U.S. military site for armored vehicles into a landscaped park and recreation area that is a habitat for many different species as well as a place for sports and recreational activities. The BUGA Park will continue to exist as a fresh air supply to cool the city center by an average of 1.5 degrees and as a public park to enhance the quality of life for Mannheim's citizens. At BUGA, a wide variety of links are created and related to the 17 SDGs adopted by the UN. In addition to the unsealing of 80 hectares of land, the cooling down of Mannheim's City Center by up to 1.5 degrees and the expected visitor numbers of 2 million, 5,000 events related to the SDGs, New Flats for 4,500 people with 30% quota emerging species, biodiversity rises are also surveyed and the clean energy production by solar panels and new quickly attainable recreation area for 40,000 people in the surrounding living areas.

For the implementation of the BUGA, a non-profit limited liability company was founded, which worked closely with politicians, the city's urban planners, landscape architects, citizens, initiatives, and many other stakeholders. Several participation processes were carried out for the development of the Federal Gardening Show and results were presented in the process in the neighborhood. During this time, the exhibition area serves as a stage for associations, government, and initiatives of various topics, all related to the SDGs with at least 5000 events. Examples of events today: "New Mobility for Families"; "Climate neutral Mobility-Flying with fuel cell"; "Insect friendly green care".

A non-profit company with temporarily hired 65 people was founded in 2014 for the preparation and implementation of the Federal Gardening Show Mannheim 2023, which will be dissolved after the event and will then be transferred to the City of Mannheim. Investment in the transformation of military land (contaminated site disposal, unsealing, etc.) into ecological green space is about 100 million euros.

With the BUGA, Mannheim has established a format to actively inform residents and visitors about the SDGs and encourage people to act. The city is still in a movement and the BUGA and the SDG as a tool catalyzed its ambitions, so that it now—what was not seen be able three years ago—proclaimed the Goal to be CO_2 neutral by 2030, which leads the city to be officially accepted as one of European 100 Climate Neutral Cities in Europe.

The city faced obstacles in transforming the area during the COVID-19 pandemic, where a lot of work was not possible. Also with the Refugee crisis, there were a lot of refugees temporarily located in the former military buildings and sometimes it was not clear whether the Federal Government would use this area permanently as a refugee camp. Unexpected contaminated sites remained as the City's unexpected political citizens initiatives with "unconvertible" ideas about the contaminated areas and a view of financing the whole, which was not realistic in any way. So the City discussed a lot and found a way to convince most of them.

The BUGA site not only brings fresh air from the mountains into the city center but cools it down by up to 1.5 degrees. In addition to providing local recreation opportunities for citizens and especially for families, it also offers habitat for many protected and rare animal and plant species, this is a new, green area with 80 hectares in the city. Social housing could be created on part of former military areas. In addition, a new bicycle expressway has been created, which is driving the mobility turnaround in Mannheim.

The city is using the document to examine how to develop and green former U.S. military land (German Federal Garden Show). From 2024, the BUGA will be a recreational area and meeting place for the entire city and especially its neighboring areas. As an example of work, Mannheim created a park that provides a fresh air corridor to the city center, reducing temperatures by 1.5 °C in the summer (Federal Garden Show). Beneficiaries are not just Mannheim residents, but also inhabitants of the metropolitan region (2.3 million). And now there are 2 million visitors expected for the BUGA year.

Summary of Lessons Learned for the Second Part of the Project

Globally, the COVID-19 pandemic has demonstrated the importance of recreational areas with green space, fresh air, and opportunities for exercise, especially in urban areas, especially during times of lockdown. The unsealing of an 80-hectare area, which provides habitat for a wide variety of plant and animal species, has created a space for this. Children's playgrounds for all ages, opportunities for exercise for cyclists, track athletes, roller skaters, joggers and many more are of paramount importance for the health and well-being of urban citizens.

Related SDGs: Goals 3, 11, and 13.

Acknowledgements We explicitly thank the project owners based in Mannheim, Germany, for their participation and the provision of materials for the 6th Guangzhou International Award for Urban Innovation. We also congratulate them on their achievements in the 2023 cycle of awards.

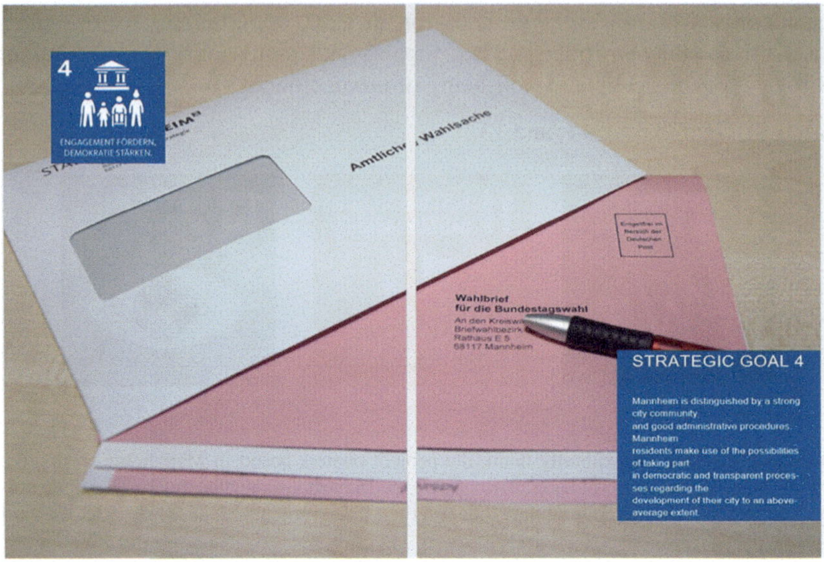
4
ENGAGEMENT FÖRDERN, DEMOKRATIE STÄRKEN.
Amtliche Wahlsache
Wahlbrief
für die Bundestagswahl
STRATEGIC GOAL 4
Mannheim is distinguished by a strong city community and good administrative procedures. Mannheim residents make use of the possibilities of taking part in democratic and transparent processes regarding the development of their city to an above-average extent.

5
INNOVATIONEN VORANTREIBEN, TALENTE GEWINNEN.
STRATEGIC GOAL 5
As a digital and innovative metropolis, Mannheim creates the conditions for companies of every size to realize diverse and sustainable growth as well as attract talented and skilled employees.

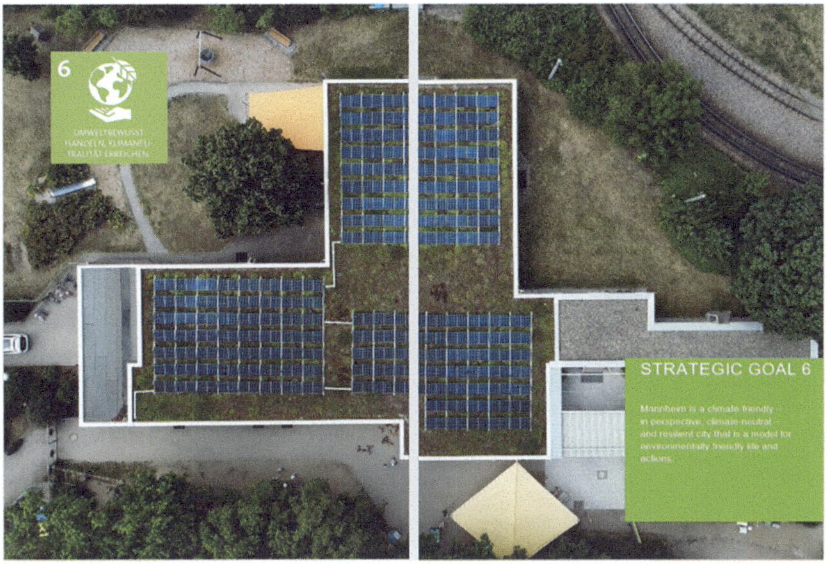

The above images are provided to us by the project owners for their submission.

Chapter 12
Navi Disha: Community Toilet Model in Pimpri Chinchwad, India

Abstract Originally run by outside organizations, the Navi Disha project was started to address issues with the upkeep and operation of 160 communal restrooms spread across 71 slums. The municipal corporation implemented a community-driven strategy under the Swachhagrah mission, giving women's self-help groups (SHGs) the authority to manage these facilities. In addition to ensuring better sanitation, this bottom-up approach promotes community ownership and gender empowerment. Navi Disha fosters long-term sustainability by actively integrating beneficiaries in the management process, offering hygienic and secure sanitation while enhancing local accountability and involvement. Additionally, by integrating women into the sanitation workforce, the initiative supports economic opportunities for women, promotes behavioral change, and raises awareness of hygiene. SHGs acquire the skills necessary for efficient operation and maintenance through training and capacity-building initiatives, guaranteeing that the restrooms continue to be operational and well-maintained. This model illustrates how sustainable urban development and improved service delivery can result from participatory governance. The initiative is related to SDGs 1, 3, 5, 6, 9, and 11.

Keywords Sanitation · Community engagement · Women empowerment · Sustainable development · Public health · Swachhagrah · Urban governance

12.1 Introduction to Pimpri Chinchwad, India

Situated in the western Indian state of Maharashtra, Pimpri-Chinchwad is one of the nation's fastest-growing urban centers. The city, which has a population of about 2.96 million and is growing at a fast rate of 4.63% per year, has major problems with its urban infrastructure, especially with regard to public health and sanitation. With a population density of 9,600 per square kilometer and an area of 181 square kilometers, Pimpri-Chinchwad places a tremendous strain on the sanitation infrastructure that is currently in place. Industrial development is the main driver of the city's economic

A. Cheshmehzangi et al., *People-Centered Innovation*, Urban Sustainability,
https://doi.org/10.1007/978-981-96-5230-3_12

prosperity; a large concentration of manufacturing facilities, the automobile industry, and small businesses contribute to its $3,840 GDP per capita.

Like many growing Indian cities, Pimpri-Chinchwad has experienced sanitation problems despite its industrial and economic expansion, especially in slum and informal settlement areas. Open defecation has been a major issue, with direct effects on environmental sustainability, public health, and general quality of life. The municipal corporation responded by launching extensive initiatives to increase access to sanitary facilities, emphasizing locally driven solutions to guarantee long-lasting effects.

The municipal corporation launched Navi Disha, a ground-breaking program under the Swachhagrah campaign, after realizing that conventional top-down approaches to sanitation management frequently fail because of poor maintenance and a lack of community involvement. By transferring management of the city's 160 community restrooms, which are dispersed throughout 71 slum clusters, from outside organizations to women's self-help groups (SHGs), the program sought to address operational issues. This calculated action was intended to empower women in the area while encouraging a feeling of responsibility and ownership among the communities that make use of these facilities.

The Navi Disha initiative is an example of a move toward participatory governance, where citizens and the government share responsibility for sanitation rather than it being just a city service. The program aimed to guarantee the long-term operation of these vital facilities by enlisting SHGs in the management and upkeep of communal restrooms. In order to provide safer and cleaner restrooms for women and children, who are disproportionately impacted by inadequate sanitation infrastructure, the initiative also sought to address gender-specific sanitation issues.

In addition to enhancing access to sanitary facilities, Navi Disha combines sanitation management with economic empowerment, providing women with chances to earn money by helping to maintain public facilities. In addition to improving public health outcomes, this strategy increases resilience and community cohesion. Since Pimpri-Chinchwad is a city that is closely linked to industrial growth, its achievements in sanitation reform provide a template for other quickly urbanizing areas dealing with comparable issues. Navi Disha serves as an example of how gender-inclusive policies, community involvement, and local governance can promote sustainable development while tackling pressing urban sanitation concerns. By implementing such programs, Pimpri-Chinchwad keeps moving closer to accomplishing the Sustainable Development Goals (SDGs), especially SDGs 6 (Clean Water and Sanitation) and 11 (Sustainable Cities and Communities), which guarantee that inclusive and equitable public services accompany urban growth.

12.2 Background and Origins

The urgent issues surrounding the upkeep and operation of communal restrooms in the city's impoverished neighborhoods gave rise to the Navi Disha project in Pimpri-Chinchwad. The municipal corporation made an effort to provide sanitation infrastructure, but many community restrooms were neglected, vandalized, and in unsanitary conditions. Facilities began to deteriorate due to improper maintenance, which discouraged use and forced some residents to defecate in the open. Concerns regarding women's and girls' safety, especially at night and in the early morning, also brought attention to the need for a more secure and inclusive approach to sanitation management.

Navi Disha, a community-driven project that aims to provide everyone with safe, hygienic, and easily accessible sanitation facilities, was started by the Pimpri-Chinchwad Municipal Corporation (PCMC) in response to these issues. Along with addressing important goals of SDG 1 (No Poverty), SDG 3 (Good Health and Well-Being), SDG 5 (Gender Equality), SDG 11 (Sustainable Cities and Communities), and SDG 16 (Peace, Justice, and Strong Institutions), the initiative supports Sustainable Development Goal (SDG) 6 (Clean Water and Sanitation). Navi Disha aims to deliver essential services while reducing the risks associated with poor sanitation by combining sanitation improvements with more comprehensive urban resilience strategies described in the New Urban Agenda.

The initial hesitancy and lack of confidence among the community groups tasked with maintaining the restrooms was one of the main obstacles to project implementation. Numerous women's self-help groups (SHGs), which were recognized as important initiative participants, were reluctant to assume the duty because they lacked facility management expertise. In order to get around this, PCMC implemented a training-of-trainers model, giving women the tools they needed to effectively run and maintain the restrooms. These women groups were trained and mentored by a group of 17 social coordinators, sanitary inspectors, and senior municipal officials.

The project effectively enabled local women to take charge of sanitation facilities through organized training programs and ongoing handholding support, encouraging a sense of accountability and responsibility among the community. By addressing problems like facility misuse and encouraging hygienic practices, the skilled social coordinators also sought to maintain a high level of community involvement. Navi Disha has turned public toilet management from a municipal problem into a community-led success story by fusing sanitation with gender empowerment and participatory governance. This shows how local governments can collaborate with citizens to develop inclusive and sustainable urban solutions.

12.2.1 Key Innovative Aspects

The Navi Disha initiative's strong focus on gender inclusion, which places women at the forefront of community sanitation management, is one of its key innovations. Navi Disha gives women's self-help groups (SHGs) the authority to take control of public restrooms, in contrast to traditional municipal sanitation programs that frequently depend on outside contractors or male-dominated workforce structures. Beyond merely offering essential services, this strategy gives women access to economic and social opportunities that allow them to grow as leaders, become financially independent, and gain influence in their communities.

Navi Disha makes sure that the special needs of women and girls—especially those related to safety and hygiene—are given top priority in sanitation planning and management by promoting a gender-responsive governance model. The initiative acknowledges that creating resilient and inclusive communities where women actively participate in decision-making and service delivery is just as important to sustainable urban development as improving infrastructure.

Navi Disha's agile administration model, which enables constant strategy adaptation and improvement for increased efficacy and efficiency, is another distinctive feature. Real-time learnings and feedback mechanisms have been incorporated into the initiative to address new challenges as it has developed dynamically. For instance, modifications to remuneration models were implemented to guarantee equitable and prompt monetary compensation for women in charge of the restrooms, encouraging continued involvement. Structured monitoring frameworks were also put in place to keep tabs on performance, spot gaps, and offer the required interventions. In order to guarantee women groups' sustained participation, the program also gave special attention to capacity-building and motivation, providing them with continual training, mentoring, and recognition. This adaptable and responsive governance model establishes a standard for how cities can successfully incorporate community involvement, gender empowerment, and adaptive governance into the provision of urban services.

12.2.2 Desired Outcome and Measurement

12.2.2.1 Desired Outcome

Establishing sanitary facilities in Pimpri Chinchwad that are safe, clean, and sustainable while also encouraging gender inclusion, community involvement, and economic empowerment is the clear and multifaceted goal of the Navi Disha initiative. The project's main goal is to change how community restrooms are managed and maintained by giving women's self-help groups (SHGs) ownership. This will guarantee everyone's safety, dignity, and accessibility, especially for women and underrepresented groups. The project seeks to create a long-term sustainable sanitation model that encompasses social transformation and empowerment in addition to infrastructure by strengthening the partnership between the community and the urban local body (PCMC).

Creating a sense of shared accountability and ownership for community sanitation facilities is one of Navi Disha's main goals. Public restrooms in low-income neighborhoods have historically experienced poor maintenance, vandalism, and neglect, creating unsanitary conditions and preventing vulnerable groups from using them. Navi Disha seeks to end this cycle by putting women at the center of operations and making sure that restrooms are safe, secure, and well-maintained for all. In addition to greatly lowering vandalism and facility misuse, women-led management gives them a sense of accountability and gives them the chance to gain financial stability and leadership experience through structured compensation.

Enhancing gender security by providing women with well-maintained restrooms, even after hours, is another crucial result. Women are forced to use open areas or travel great distances for basic hygiene needs in many urban slums due to the lack of safe sanitation facilities, which puts them at risk for health problems and safety issues. This problem is directly addressed by Navi Disha's strategy of women-led restroom operations, which guarantees that these areas continue to be secure, well-lit, and operational. In addition to enhancing health and hygiene, this action supports a larger cultural change that values and prioritizes women's safety and dignity in urban planning.

Beyond sanitation, Navi Disha sees a time when women have more power than just running and maintaining restrooms. The program aims to increase their self-esteem, improve their financial independence, and create a robust network of support among women in the community. Women can become leaders and role models by participating in sanitation management through Navi Disha, which opens new careers, social mobility, and civic engagement opportunities for them. The program's goal is to have a cascading effect, empowering women in fields like public health advocacy, community leadership, and entrepreneurship in addition to sanitation.

To measure the success of Navi Disha, the Pimpri Chinchwad Municipal Corporation (PCMC) has established a comprehensive monitoring framework. Key indicators include (Fig. 12.1):

- Improvement in sanitation facility maintenance, assessed through periodic quality checks and user feedback surveys.
- Reduction in vandalism and toilet misuse, tracked through maintenance reports and incident logs.
- Increased women's participation in governance, measured by the number of women actively engaged in managing facilities and leading community discussions.
- Impact on public health, evaluated through data on reduced incidences of sanitation-related diseases in the areas covered by Navi Disha.
- Social and economic impact, monitored through indicators such as income generation among women SHGs, job retention, and expansion of peer-support networks.

Fig. 12.1 Summary of key indicators for the project

Additionally, PCMC is dedicated to incorporating these women into other municipal development projects, promoting cross-learning initiatives among communities, and integrating them as peer educators in sanitation awareness programs. The city hopes to replicate and scale up this model across various sectors by establishing them as important stakeholders in urban governance, guaranteeing that women's leadership in public services becomes a long-lasting and institutionalized practice.

By implementing Navi Disha, PCMC is not only tackling sanitation issues but also advancing a comprehensive urban development model that incorporates economic inclusion, public health, and gender empowerment, illustrating how localized initiatives can support more general Sustainable Development Goals (SDGs).

12.2.2.2 Measurement

To guarantee its efficacy in empowering women-led self-help groups (SHGs) and delivering sustainable sanitation, the Navi Disha initiative uses a strong and organized measurement framework. The effective operation and upkeep of public restrooms is essential to the project's success, so the monitoring system incorporates both qualitative and quantitative measurement tools. These tools monitor the initiative's socioeconomic effects on the participating women as well as performance, cleanliness, and user satisfaction.

SHGs are responsible for keeping thorough monthly utilization records, which is a crucial part of the measurement process. Documenting the use of cleaning supplies, master cleaning plans, and user input is necessary for each group to guarantee that hygienic requirements are continuously fulfilled. These documents offer valuable insights into day-to-day operations, assisting in pinpointing areas that might benefit from extra resources or training. Since the corresponding zonal offices only approve monthly sanitation operations invoices following an on-site verification of these registers, the financial component of this monitoring system is closely related to it. This measure guarantees openness, avoids inconsistencies, and strengthens responsibility among the management teams led by women.

PCMC has assigned social coordinators to regularly inspect the communal restrooms in addition to SHG-led monitoring. These coordinators have two responsibilities: they evaluate the standards for cleanliness and upkeep and provide training to increase the proficiency of female operators. Social coordinators assist in addressing issues in real time by actively interacting with the women's groups and offering mentorship and support on the ground to guarantee the initiative's sustainability. Their recurring visits guarantee that any problems pertaining to facility administration, community involvement, or safety issues are promptly resolved, improving the project's overall efficacy.

The implementation of an annual competition among the Navi Disha community restrooms is a distinctive and inspiring feature of the measurement framework. In addition to guaranteeing excellent performance across all participating facilities, this initiative fosters a spirit of healthy competition and pride among the women-led groups. The competition's evaluation criteria include community engagement levels, user satisfaction, maintenance efficiency, and cleanliness standards. Establishments that continuously exhibit operational excellence are acknowledged and rewarded, which further boosts the female operators' motivation and dedication over the long run. This strategy fosters a culture of excellence and shared responsibility within the community in addition to reinforcing accountability.

To evaluate the initiative's impact from the viewpoint of the community members who depend on these facilities, user feedback is another crucial component of measurement. The program makes sure it stays responsive to community needs by monitoring usability issues, safety concerns, and satisfaction levels. In order to make well-informed decisions, such as boosting security, upgrading infrastructure, or modifying operational plans in light of actual usage trends, PCMC actively gathers and examines this data. PCMC's integrated approach to monitoring and evaluation ensures that Navi Disha is not just a sanitation initiative but a transformative project that promotes gender equity, economic empowerment, and sustainable urban sanitation solutions. By emphasizing transparency, accountability, and continuous improvement, the measurement framework ensures that the initiative remains impactful, scalable, and adaptable to evolving urban challenges.

12.2.3 Implementation

12.2.3.1 Principal Parties and Partners

In order to ensure that local stakeholders, especially women's self-help groups (SHGs), play a crucial role in sanitation management, the Navi Disha initiative is being implemented using a community-led model. As the knowledge partner, PCMC's City Transformation Office offers strategic direction and makes sure the project is in line with more general urban development objectives. With assistance from the Health Department and Zonal Offices, the Social Development Department oversees the initiative's governance, guaranteeing a coordinated and cooperative approach to urban sanitation.

The collaboration between PCMC and the women-led SHGs forms the basis of the implementation strategy. Building and maintaining restroom infrastructure, providing sanitary and safe cleaning supplies, and managing the overall sanitation plan are all part of PCMC's responsibilities. In turn, the women's organizations oversee operations and maintenance (O&M), raising awareness and promoting behavioral change in their local communities. This model ensures the initiative's sustainability by promoting community ownership in addition to raising sanitation standards. The Zonal Offices are essential to daily oversight and financial management. Based on documented records of toilet usage, cleanliness, and maintenance, they guarantee that monthly operating costs are met. A systematic evaluation system has been put in place to ensure efficiency, and PCMC officials conduct routine facility inspections, evaluate hygienic standards compliance, and give SHGs feedback. This ongoing involvement guarantees that female operators get the resources, assistance, and training they need to run the facilities efficiently.

The success of the initiative is largely dependent on the participation of women's organizations as key partners. These women acquire leadership abilities, financial literacy, and technical knowledge through capacity-building initiatives, which equips them to manage sanitation facilities effectively. A notable change in sanitation practices, including a decrease in vandalism and an encouragement of responsible use of public restrooms, has also resulted from their involvement in awareness campaigns and community engagement initiatives. The program supports gender-inclusive urban development and helps women become economically independent by empowering them as sanitation entrepreneurs.

The initiative is strengthened by external collaborations in addition to local governance. PCMC's City Transformation Office offers proficiency in impact assessment, creative urban solutions, and policy alignment. Additionally, a number of non-profits and social organizations support knowledge-sharing among various women-led groups, facilitate peer-learning exchanges, and conduct training workshops. These collaborations guarantee that Navi Disha stays flexible in the face of new difficulties and aid in scaling up best practices. In addition to being an infrastructure project, the Navi Disha initiative is a transformative model of inclusive and sustainable sanitation governance thanks to the comprehensive cooperation between PCMC, women's organizations, government agencies, and knowledge partners. The project creates the foundation for long-term effects by fusing community ownership with policy-driven governance, improving urban sanitation and socioeconomic development.

12.2.3.2 Resources for Implementation

Strong contractual and operational resources that are precisely measured to guarantee the sustainability and caliber of community sanitation facilities form the basis of the Navi Disha project's execution. PCMC has set up a contract under the project that requires the managing group to receive INR 900 (roughly USD 11) per toilet seat on a monthly basis. In order to guarantee that the community restrooms are kept in excellent condition without financial limitations impeding day-to-day operations, this sum has been carefully calculated to cover necessary cleaning supplies like brooms, floor wipers, buckets, mugs, and other miscellaneous expenses.

The municipal corporation actively contributes to the provision of safe consumables required for sanitation in addition to the financial arrangements. PCMC provides buckets, soaps, disinfectants, hand gloves, and other cleaning supplies directly to the women-led self-help groups in charge of running and maintaining these facilities. By allocating resources proactively, the sanitation teams can maintain hygiene standards even in the face of high usage and possible wear and tear without being hampered by a lack of supplies.

The project's focus on improving record-keeping capacity is a key component of its resource management. The women's organizations are responsible for keeping thorough monthly utilization records. These records include monitoring the use of cleaning supplies, making sure master cleaning schedules are followed, and getting user input. Because they form the foundation for financial reconciliations and performance reviews, the quality and accuracy of these records are critical. By ensuring that the monthly invoices are only paid after the corresponding zonal offices have confirmed the information through on-site visits to the registers, PCMC fosters an environment of transparency and accountability. In order to help the groups even more, PCMC has trained 17 social coordinators who offer continuing technical support and mentoring. By making sure that best practices in sanitation management are followed and that any new problems are dealt with right away, these coordinators play a crucial role in bolstering the operational capabilities of the women-led teams. In addition to promoting ongoing capacity building among the female operators, their frequent visits to the communal restrooms aid in maintaining cleanliness and operational effectiveness.

The project uses a novel incentive system in an effort to inspire and maintain high performance. A half-yearly incentive in the form of a USD 60 flexi-fund is available to groups that maintain a record of zero vandalism. The purpose of this incentive is to promote timely repairs, proactive maintenance, and the development of creative ideas for improving facility management. The flexi-fund ensures that the sanitary facilities are not only kept up to date but also constantly changing to satisfy the needs of the community by serving as a reward for hard work and a spur for additional advancements. The foundation of Navi Disha is made up of these carefully considered resources and implementation techniques. Through the integration of financial assistance, consistent consumable supply, meticulous documentation, committed training, and performance-based rewards, the project establishes a strong foundation for sustainable urban sanitation. In addition to ensuring that the restrooms stay safe and clean, this all-encompassing strategy empowers local women, encourages community ownership, and establishes a model for efficient public service delivery in urban settings.

12.2.3.3 Tools and Methods Applied

The Navi Disha project in Pimpri Chinchwad sought to address the sanitation issues faced by the urban poor by developing a community-led, sustainable toilet model. A wide range of instruments and techniques were used to accomplish this, guaranteeing a long-term effect and a participatory approach. One of the main components of the community engagement process was the use of multiple Focus Group Discussions (FGDs). Regular focus group discussions (FGDs) with local stakeholders, residents, and community representatives helped the project team comprehend the concerns, preferences, and sanitation needs of various groups. Residents were able to express their experiences with the sanitation infrastructure and their future goals through these conversations. In order to design the restrooms to be more inclusive, accessible, and in line with the social and cultural norms of the community, the input received was crucial.

To appraise the current sanitation facilities, find any gaps, and determine whether the infrastructure is adequate, surveys of the infrastructure were carried out. To comprehend the difficulties with accessibility, cleanliness, and upkeep, thorough surveys of public restrooms, drainage systems, and waste disposal techniques were conducted. In order to ensure that the new community restrooms were positioned and furnished to suit the needs of the local populace, this data was essential to their design. Pimpri Chinchwad's sanitation procedures were compared to best practices from other cities in India and abroad using benchmarking studies. These studies offered insightful information about creative and long-lasting urban sanitation solutions. Understanding what had worked in other areas allowed the project team to modify their strategy to fit the local setting, adding design features that had worked well in other places while modifying them to meet Pimpri Chinchwad's unique requirements.

To raise community awareness of the value of good sanitation and hygiene habits, awareness campaigns were held. To disseminate important messages, these campaigns made use of a variety of communication tools, such as workshops, posters, and local media. Involving community leaders and local influencers helped spread the word and encouraged a sense of accountability and ownership among the public. Lastly, perception analysis was used to track how the community felt about the new restrooms over time. Measures of satisfaction and areas for improvement were determined with the aid of surveys and follow-up interviews. This analysis made sure the project was adaptable and responsive, enabling changes to be made as necessary to sustain community involvement and support. When combined, these resources helped create a flexible, open, and community-driven framework that was essential to the Navi Disha project's success in enhancing sanitation in Pimpri Chinchwad.

12.3 Obstacles and Solutions

When putting the Navi Disha Community Toilet Model into practice, the Pimpri Chinchwad Municipal Corporation (PCMC) encountered a number of noteworthy challenges, especially when it came to encouraging a sense of community ownership. Building toilets was only one aspect of the problem; another was making sure they were used, maintained, and incorporated into daily life. One of the main challenges was encouraging community responsibility, especially among women, for the responsible use of public restrooms. Several rounds of community consultations were conducted to address this, with women in particular showing a strong interest in the project. As a result, the Navi Disha model was created, putting women at the center of the solution.

While the project made strides in building an institutional framework, PCMC encountered several challenges, each requiring targeted solutions. Below is a breakdown of these challenges and the solutions that were implemented to overcome them:

Challenges Addressed by PCMC:

1. Lack of Family and Community Support for Women's Groups and the Initiative: One of the first hurdles was the limited support from families and the broader community towards the women groups involved in managing the toilets. The traditional mindset regarding women's roles in sanitation was an obstacle, and without strong community backing, the sustainability of the initiative was at risk.
2. User Group Behaviour and Vandalism: The issue of vandalism was a significant challenge, with toilets being misused or damaged by the very community they were meant to serve. This behaviour was often linked to a lack of respect for the infrastructure and an absence of community accountability.
3. Lack of Robust Monitoring Framework: The operation and maintenance of community toilets lacked a structured, consistent monitoring system. Without effective oversight, there was a risk of toilets falling into disrepair, thereby negating the project's long-term sustainability.

4. Lack of Institutionalization: The project lacked a formalized structure for its operations and governance, making it difficult to scale the model or ensure its integration into the city's formal sanitation system.

Solutions Undertaken by PCMC:

1. Building Capacity and Formalizing Women's Groups: To empower the women managing the toilets, PCMC organized capacity-building sessions, helping women gain essential skills and knowledge for managing and maintaining community toilets. These sessions also focused on formalizing the women's groups, making the process of registration easier and more accessible, thereby enhancing the credibility and autonomy of these groups.
2. Information, Education, and Communication (IEC) Activities: A series of IEC campaigns were conducted in community hotspot areas to raise awareness about the importance of proper sanitation, the usage of community toilets, and behaviour change. These activities, led by the women groups themselves, also aimed at educating the public to prevent acts of vandalism and misuse. These campaigns played a critical role in changing community behaviour, fostering greater respect for the toilets, and encouraging responsible usage.
3. Formalizing the Strategy and Change Processes: PCMC implemented a formalized strategy for the operation and maintenance of the toilets, including clear guidelines for the management of toilets, monitoring, and resolving any issues. This approach helped establish a more transparent, accountable system that could be more easily scaled and replicated in other parts of the city.
4. Training of Trainers to Scale the Model: To expedite the scaling of the Navi Disha model, PCMC trained trainers from various levels of administration, including Social Coordinators, Area Health Officers, and Regional Officers. These trained personnel were tasked with implementing the model across 40 community toilets throughout the city, ensuring that the initiative reached a wider population and had greater impact.
5. Family and Community Capacity Building: In addition to focusing on the women groups, PCMC also conducted several capacity-building activities targeting their families and the wider community. These activities were crucial in creating a supportive environment for the women and fostering greater community engagement in the project. By educating families and local stakeholders, the initiative built broader social support, which was essential for the success and sustainability of the model.
6. Firm Administration Support in Addressing Vandalism: The city administration took a strong stance in supporting the women's groups in cases of vandalism and misuse of infrastructure. The administration provided the necessary backing to resolve conflicts, and through the ongoing IEC campaigns, local communities were encouraged to take ownership and help prevent future vandalism. This collaboration between the local administration and the women groups was pivotal in ensuring the protection and maintenance of the toilets.

By using these solutions, the PCMC was able to overcome the obstacles the project faced and develop a sustainable, community-driven model that empowered women, involved the larger community in urban development, and enhanced sanitation.

12.3.1 Achieved Outcome

The Navi Disha initiative has significantly improved sanitation and empowered women in Pimpri Chinchwad, as evidenced by a number of noteworthy results. Among the most noteworthy accomplishments is the active participation of more than 400 women, who are currently in charge of overseeing 40 communal restrooms with a combined capacity of 862 toilet seats. In addition to meeting a pressing community need, this program has improved public health and hygiene for thousands of locals by giving over 30,000 beneficiaries daily access to safe and clean restrooms. Women's involvement in this process has improved their social security and general well-being by addressing sanitation issues and empowering them economically.

These women's organizations have benefited from their participation by becoming more respected and well-known in their communities in addition to becoming financially independent. As they oversee the upkeep, cleanliness, and operation of the public restrooms, they have evolved into essential change agents. A strong sense of accountability and ownership has resulted from this, which is essential to the initiative's long-term viability. The women's participation has also changed how sanitation is viewed and managed locally, which has had a wider social impact.

Since the model's success has spread throughout the community, other women's organizations have expressed immense interest in taking over the management of Pimpri Chinchwad's remaining public restrooms. This increasing demand from neighborhood women's organizations demonstrates both the Navi Disha model's effectiveness and the desire for more participation in public service projects. In light of this change, PCMC has made the decision to expand the project, intending to involve more than 1,200 women and expand the model to include more than 120 public restrooms. More than 2,062 toilet seats will be added as a result of this expansion, greatly expanding the project's capacity and reach.

PCMC aims to increase the number of impacted beneficiaries by at least three times in accordance with the scale-up. This objective demonstrates the dedication to making sure that more locals, particularly those from underserved communities, have steady and dependable access to sanitary facilities. The city's social fabric will be strengthened, and women's financial situation will improve if they continue to manage these restrooms. The project's expansion also helps Pimpri Chinchwad's general urban sanitation, which is in line with more general objectives for community well-being and urban sustainability. The Navi Disha model's success in Pimpri Chinchwad serves as an example of how community-driven projects, especially those that empower women, can result in both infrastructure and social change. With the project's expansion, there is still a great chance for additional impact and long-lasting change that will help future generations.

12.4 Lessons Learned

To conclude, the Navi Disha project in Pimpri Chinchwad, India, has shown the ability of community-driven models to revolutionize urban sanitation systems, making it a ground-breaking endeavor. The project has taught us a lot about how well local knowledge, community involvement, and gender empowerment can be incorporated into urban development. In addition to improving public sanitation, the Pimpri Chinchwad Municipal Corporation (PCMC) has improved the economic and social circumstances of women in the region by switching from a traditional vendor-based sanitation model to a community-led framework.

Building community capacity is crucial for ensuring the sustainability of such initiatives, which is one of the project's most important lessons. By collaborating closely with the women's organizations in charge of running the restrooms, PCMC was able to successfully facilitate this transition. Capacity-building sessions led by Social Development department officers and technical consultants were crucial in providing these women with the skills they needed to oversee and maintain the communal restrooms. In order to guarantee that the restrooms were not only kept up effectively but also ran in a way that benefited the whole community, these sessions also aimed to equip the women with knowledge of financial management, operational best practices, and regulatory procedures. Officials' active participation and the PCMC administration's seamless transition to this change made sure that operations continued as planned, highlighting the significance of institutional dedication to community-driven initiatives.

The successful convergence of several dimensions—economic development, women's empowerment, and health improvement—is another important lesson to be learned from the Navi Disha project. The program offered beneficiaries comprehensive support, from training and capacity building to regulatory advice and help formalizing women's organizations. By taking a comprehensive approach, the project was able to address the long-term financial security of the women in charge of the restrooms as well as the urgent need for safe and clean sanitation. The project also promoted a sense of pride and ownership by providing these women with the means to oversee the infrastructure, which was crucial to its long-term success. Additionally, the implementation of Standard Operating Procedures (SOPs) and the creation of a monitoring framework made sure that the project remained transparent, accountable, and successful in achieving its objectives. Further illustrating the connection between sanitation, health, and economic development, the emphasis on keeping restrooms clean directly improved public health, hygiene, and general well-being, all of which had a favorable effect on the community's financial status.

Another important lesson was the project's emphasis on using local knowledge to give women opportunities to generate income. In addition to improving sanitation standards, Navi Disha empowered women by allowing them to participate in management and decision-making processes. This gave them a chance to assume leadership positions and support the growth of their communities. By ensuring that women had equal access to opportunities, this emphasis on gender equality promoted more inclusive involvement in the process of urban development. This project's success has demonstrated the transformative and long-lasting potential of community-led models that emphasize empowerment and inclusivity.

Notable are the Navi Disha model's scalability and reproducibility. More than 400 women have already benefited from the project, which has also given more than 30,000 recipients access to sanitary facilities that are safe and clean. This model's success is demonstrated by its spread throughout the city and the interest it has sparked in other women's organizations to assume management positions. Navi Disha is a replicable model for other cities facing comparable urban sanitation issues, making it more than just a local success story. It exemplifies how community ownership, creativity, and teamwork can lead to sustainable urban development. This project demonstrates how, with the correct assistance, urban sanitation can serve as a driving force behind more extensive social, economic, and environmental transformation, empowering those who are typically left out of the development process and benefiting entire communities.

Related SDGs: Goals 1, 3, 5, 6, 9, and 11.

Acknowledgements We explicitly thank the project owners based in Pimpri Chinchwad, India, for their participation and the provision of materials for the 6th Guangzhou International Award for Urban Innovation. We also congratulate them on their achievements in the 2023 cycle of awards.

नवी दिशा
Vidhisha Self Help Group E Zone,
Bopakhel Ramnagar, PCMC
Lat long = 18.587106084610856

The above images are provided to us by the project owners for their submission.

Chapter 13
Active Aging for a Vibrant Future: Empowering Seniors to Thrive Through Voluntarism, Lifelong Learning, Wellbeing, and Community-Led Initiatives in Ramallah, Palestine

Abstract The "Forum of Expertise" is a cutting-edge community center that is part of the Ramallah Municipality's initiative to promote well-being and active aging in an urban setting. By improving their quality of life, building resilience, and advancing general well-being, this program aims to improve the lives of senior men and women. The Forum of Expertise addresses social isolation and economic vulnerabilities while fostering an inclusive environment that enables older adults to continue being engaged members of society. A cross-generational platform that promotes mentorship and knowledge sharing; a structured voluntary engagement channel that allows older people to make significant contributions to community development; a focused capacity-building program that aims to improve skills and opportunities for lifelong learning; and a holistic well-being framework that takes into account physical, mental, and social health are the initiative's four main programs. Together, these elements create a vibrant atmosphere that supports the well-being of senior citizens. In addition, the Forum of Expertise supports neighborhood-based programs that promote mental, social, and psychological health, enhancing intergenerational relationships and creating a strong sense of community. The initiative builds a sustainable model that benefits all facets of society by emphasizing inclusivity and making sure that the needs of vulnerable and marginalized groups are met. This strengthens social cohesion and long-term community resilience. The initiative is related to SDGs 3 and 10.

Keywords Active aging · Community engagement · Well-being · Intergenerational exchange · Social resilience · Inclusive urban development · Elderly empowerment

13.1 Introduction to Ramallah, Palestine

An important urban hub in Palestine, Ramallah is a city rich in culture, history, and economic vitality. The City is located in the West Bank as part of Palestine. With a population of about 85,000 and a consistent yearly growth rate of 2.00%, Ramallah

A. Cheshmehzangi et al., *People-Centered Innovation*, Urban Sustainability,
https://doi.org/10.1007/978-981-96-5230-3_13

has become a significant center for commerce, administration, and culture. With 4474 inhabitants per square kilometer, the city has a comparatively high population density given its 19 km^2 size. Ramallah has proven resilient in the face of persistent regional difficulties, upholding its reputation as a hub for innovation, business, and governance.

The population of Ramallah is youthful and vibrant, and a sizable fraction works in trade, services, and tourism, the main drivers of the city's prosperity. With a GDP per capita of $3789.3, the city's economic activity is moderate when compared to the Palestinian economy as a whole. Ramallah's income distribution is comparatively balanced, as indicated by its GINI index of 0.33; however, inequalities still exist, especially among marginalized and vulnerable communities. Ramallah, one of the West Bank's most urbanized regions, has seen significant growth in recent years thanks to a mix of foreign involvement and domestic investment. The city's varied economic landscape is influenced by a variety of domestic companies, foreign organizations, and non-governmental organizations. Ramallah's standing as a hub of political and economic power has been further cemented by the influx of professionals, businesspeople, and foreigners.

Ramallah's economy is heavily reliant on trade and tourism. With a large number of markets, shopping malls, and service-oriented enterprises serving both locals and tourists, the city boasts a flourishing commercial sector. Despite being influenced by political ups and downs, the tourism sector continues to be a vital part of the economy, bringing tourists to Ramallah's museums, cultural sites, and lively social scene. In addition to being the center of trade, Ramallah houses important government offices, international organizations, and foreign consulates, making it the de facto administrative capital of the Palestinian Authority. Local economic activity has been further stimulated by the steady influx of professionals and public sector workers brought about by this administrative function. The city has also witnessed an increase in media, finance, and technology-based businesses, which reflects a changing economic environment that combines established industries with new ones.

Ramallah has experienced substantial urban growth despite its small size, with new mixed-use, residential, and commercial projects changing the cityscape. In order to make the city more livable, investments have been made in green areas, public services, and transportation networks in recent years. But there are still issues, especially regarding affordable housing, traffic congestion, and fair access to public services. In order to ensure that new construction blends in perfectly with the city's rich architectural legacy, urban planning initiatives in Ramallah aim to strike a balance between modernization and historical preservation. In order to preserve the social fabric and encourage interaction between various groups in the city, public areas, cultural institutions, and community centers are essential.

Ramallah is well known for its thriving cultural scene, which includes a number of festivals, art shows, and literary gatherings all year long. The city's reputation as a center of creativity and intellectual exchange is further supported by the artistic expression offered by its theaters, galleries, and cultural institutions. The city is also renowned for its varied culinary scene, which combines modern international influences with traditional Palestinian cuisine. A hallmark of Ramallah life is community involvement, with vibrant civil society organizations actively advancing human rights, social cohesion, and sustainable development. Local efforts, which are firmly anchored in Palestinian cultural and historical identity, emphasize environmental sustainability, women's rights, and youth empowerment.

Ramallah's development as an urban center shows how resilient and flexible it is in the face of political and economic upheaval. The city maintains a careful balance between tradition and modernization as it develops further as a center for governance, trade, tourism, and cross-cultural interaction. Ramallah has significant obstacles as it grows in the areas of social justice, infrastructure, and economic diversification. Nonetheless, Ramallah continues to be a vital pillar in the larger Palestinian urban landscape, providing a distinctive model of development and resilience thanks to its rich cultural legacy, involved citizens, and calculated economic initiatives.

13.2 Background and Origins

The Ramallah Municipality's historic initiative to create the "Forum of Expertise" demonstrates its dedication to promoting social cohesion, inclusivity, and active aging. The municipality has taken the initiative to guarantee that no member of the community is excluded because of age-related limitations because it recognizes the priceless contributions made by older generations. This program is in line with the city's larger social duty to improve the dignity and general well-being of its senior citizens.

At its core, the *Forum of Expertise* embodies the principle of *"leave no one behind,"* a fundamental value within the United Nations Sustainable Development Goals (SDGs). By championing the rights, participation, and well-being of older adults, the initiative directly supports SDG 3 (*Good Health and Well-being*), SDG 10 (*Reduced Inequalities*), and SDG 11 (*Sustainable Cities and Communities*). These goals emphasize the importance of social inclusion, lifelong learning, and accessible urban environments that cater to all age groups. The *Forum of Expertise* not only addresses immediate needs but also reinforces a long-term vision for a sustainable, age-friendly urban future.

The program also emphasizes the value of wisdom-sharing and respect between generations, which are deeply ingrained in Palestinian culture. The Forum of Expertise fosters intergenerational understanding and community ties by appreciating the wisdom and experiences of senior citizens. This local initiative makes sure that growing older is seen as a chance for ongoing involvement and contribution rather than as a sign of decline. Growing concerns about health disparities, social isolation, and decreased community participation among the elderly led to the creation of the Forum of Expertise. The World Health Organization (WHO) reports that the risks of loneliness, dependency, and restricted access to vital health and social services are rising for older adults globally. These problems are even more severe in Palestine, where the vulnerability of the elderly is increased by the lack of a comprehensive social security system and universal health coverage.

Given these obstacles, the Ramallah Municipality aimed to create a planned, neighborhood-based solution that not only takes care of urgent issues but also helps senior citizens become more resilient over the long run. In order to keep older people active, respected, and supported in their communities, the Forum of Expertise was created as a multifaceted project that incorporates social interaction, skill-sharing, and well-being initiatives. The initiative fights the detrimental effects of aging-related marginalization by establishing a venue where senior citizens can mentor younger generations, share their knowledge, and engage in worthwhile social and volunteer activities. It offers a forum for elder empowerment, recognizing them as active participants rather than passive care recipients. By encouraging intergenerational cooperation and solidarity, this proactive approach not only enhances the quality of life for the elderly but also fortifies the fabric of Palestinian society. The Forum of Expertise is ultimately a trailblazing example of age-friendly urban planning in Palestine. By emphasizing participation, respect, and ease of access, the project establishes a model for local governments in the area to follow, guaranteeing that senior citizens get the assistance, acknowledgment, and inclusion they deserve.

13.2.1 Key Innovative Aspects

By utilizing the knowledge, expertise, and life experiences of senior citizens, the Forum of Expertise offers a novel approach to active aging and fosters a vibrant, welcoming community. This initiative, in contrast to traditional elderly care programs, focuses on empowerment rather than just assistance, ensuring that older adults continue to play an active and significant role in society. The program's emphasis on intergenerational exchange, community-driven structure, and varied activity framework are its most inventive features.

The Forum of Expertise organizes a variety of events to support its members' social, intellectual, and physical health. These activities go beyond conventional leisure programs by incorporating skill-based, structured activities that foster community service and personal development (Fig. 13.1).

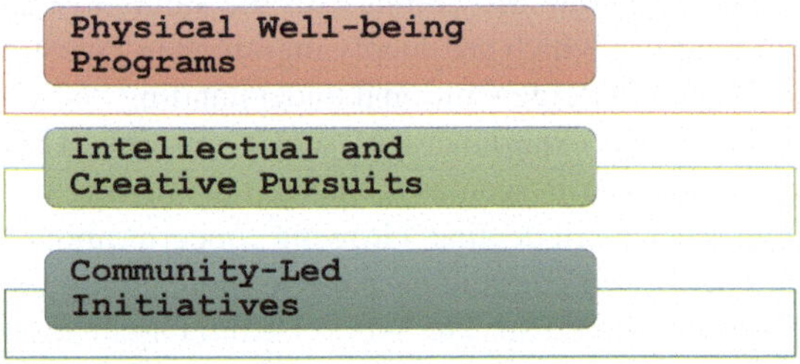

Figs. 13.1. Summary of events and programmes developed and organised under the Forum of Expertise initiative

- **Physical Well-being Programs**—The initiative offers tailored physical activities such as *chair yoga*, movement therapy, and guided exercise sessions, designed to accommodate different mobility levels while promoting flexibility, strength, and mental relaxation. These programs help combat the risks of sedentary lifestyles among the elderly.
- **Intellectual and Creative Pursuits**—Members engage in *book clubs, writing workshops, and knowledge-sharing sessions* that encourage lifelong learning and intellectual stimulation. These activities foster curiosity and creativity, preventing cognitive decline while reinforcing the value of accumulated knowledge.
- **Community-Led Initiatives**—The *Forum* supports projects initiated by its members, such as skill-sharing workshops, mentorship programs, and cultural performances. For instance, *community choirs* have been formed where elderly participants make decisions about repertoire selection and training, ensuring that artistic expression remains an integral part of aging.

This diverse programming ensures that older individuals remain actively engaged, contributing their skills while maintaining a sense of purpose and fulfillment.

The decentralized and community-driven structure of the Forum of Expertise is one of its main innovations. This initiative actively involves senior citizens in decision-making, supervision, and evaluation, in contrast to top-down models where programs are prescribed by outside organizations (Fig. 13.2).

- **Participatory Decision-Making**—Members have direct input into the planning and execution of activities, ensuring that programs align with their interests and needs.
- **Community-Led Supervision**—Rather than relying solely on municipal oversight, the initiative empowers participants to take leadership roles in organizing and moderating programs.
- **Regular Feedback Loops**—A system of continuous feedback is in place, where members evaluate the effectiveness of different programs and suggest modifications. This iterative process ensures adaptability and responsiveness to changing needs.

Decisions about training schedules, song selections, and performance arrangements, for instance, are made solely by the members of the community choir. By

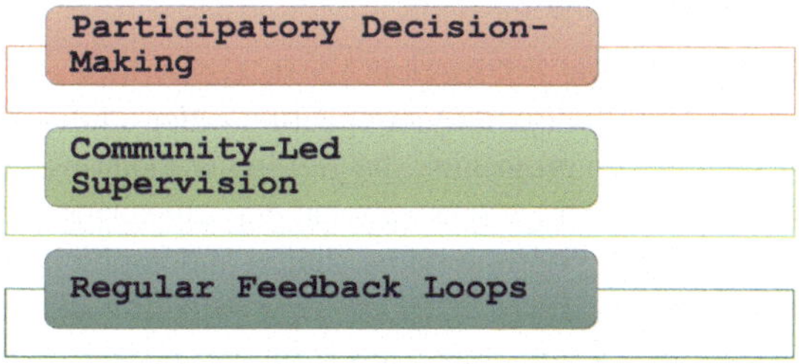

Fig. 13.2 Summary of the decentralized and community-driven structure of the Forum of Expertise

encouraging a sense of ownership, this participatory model strengthens the Forum's function as a venue for cooperation as opposed to passive participation.

By encouraging deep connections between generations and various societal sectors, the Forum of Expertise goes beyond conventional models of elder care. Through mentorship programs, where senior members impart their knowledge to younger generations in areas like entrepreneurship, storytelling, and handicrafts, it actively fosters intergenerational engagement. In order to give its members organized learning and volunteer opportunities, the initiative also works with nearby businesses, cultural organizations, and universities. Through these collaborations, senior citizens can share their knowledge and gain new insights and skills in a reciprocally beneficial setting.

The Forum of Expertise is a trailblazing approach to active aging that embraces community leadership, empowerment, and creativity in addition to traditional support networks. Its intergenerational approach, decentralized governance, and comprehensive activity framework established a new benchmark for inclusive urban aging programs. This program improves older adults' well-being and fortifies the social fabric of the larger community by emphasizing participation, respect, and lifetime contribution.

13.2.2 Desired Outcome and Measurement

13.2.2.1 Desired Outcome

By tackling the widespread problem of social isolation and promoting a more cohesive, inclusive, and active community, the Forum of Expertise aims to change the social environment for senior citizens. This program seeks to bring about a long-lasting change in societal structures, laws, and perceptions of aging rather than just a temporary fix. Fundamentally, the program aims to address social isolation and loneliness among senior citizens by offering organized chances for interaction, education, and camaraderie. The Forum facilitates meaningful relationships and a sense of purpose for older adults through interactive programs, community-led activities, and intergenerational exchanges. The goal is to create an emotionally satisfying and socially engaged atmosphere where senior citizens are respected and integrated into daily community life.

- *Encouraging Interaction Across Generations*—The initiative promotes intergenerational dialogue by connecting older adults with younger generations through mentorship programs, storytelling workshops, and cultural events. These interactions help bridge generational gaps, ensuring that the experiences and wisdom of the elderly continue to be shared and appreciated.
- *Expanding Social Circles*—By engaging in collective activities such as group exercises, book clubs, and volunteer projects, members cultivate new friendships

and strengthen their support networks. This not only enhances mental well-being but also provides a safety net for emotional support.

Achieving meaningful social inclusion for older adults requires more than just programmatic interventions—it demands a fundamental shift in the way society views and supports aging. The *Forum of Expertise* advocates for long-term societal and policy changes that prioritize senior engagement as an integral part of community development.

- *Embedding Senior Engagement in Urban Planning*—The initiative envisions a future where cities are designed with age-friendly infrastructure, ensuring that public spaces, transportation, and community centers are accessible and welcoming to older adults.
- *Influencing Policy for Greater Inclusion*—By demonstrating the benefits of active aging programs, the *Forum* aims to inspire local governments and organizations to adopt policies that integrate elderly participation in decision-making, social services, and community projects.

The *Forum of Expertise* recognizes that overcoming isolation is not a one-time effort but an ongoing process requiring sustained commitment and adaptive strategies. As social dynamics evolve, so too must the approaches used to support aging populations.

- *Continuous Improvement and Expansion*—The initiative seeks to refine and expand its offerings based on community feedback, ensuring that programs remain relevant and impactful.
- *Cultural and Behavioural Shifts*—By normalizing active aging and promoting the contributions of the elderly, the Forum aims to reshape societal attitudes, fostering a culture that values aging as a time of continued growth and participation rather than decline.

The Forum of Expertise works to establish a community in which senior citizens are respected, actively involved, and cared for. The project lays the groundwork for a more caring and age-friendly future by tackling isolation, influencing laws, and integrating social inclusion into society as a whole. The Forum aspires to a society in which all generations prosper together, cultivating a culture of meaningful connection and lifetime contribution through sustained dedication and changing tactics.

13.2.2.2 Measurement

A thorough and organized measurement framework is in place to guarantee the Forum of Expertise's efficacy and long-term viability. To monitor progress, evaluate impact, and direct ongoing improvement, the initiative uses both quantitative and qualitative indicators. The program makes sure that its goals are in line with the changing needs of senior citizens and the larger community by methodically collecting data on participation, engagement levels, and social interactions.

Senior engagement is one of the most important success factors, and it can be assessed by regularly monitoring attendance at different events and initiatives. In order to provide a more nuanced understanding of participant demographics, forum volunteers keep thorough attendance records that break down attendance by age and gender. This information aids in determining participation patterns, evaluating the initiative's outreach, and guaranteeing that a variety of senior citizen groups are reaping the benefits. Prolonged participation and high attendance rates are reliable markers of the program's value and appeal.

The number of volunteers actively participating in the Forum is another important metric. In addition to showing community support, volunteer engagement demonstrates the initiative's capacity to promote intergenerational dialogue and a supportive culture. The program measures how well it fosters cross-generational relationships by counting the number of younger volunteers, peer educators, and senior mentors who participate in various activities. In order to bridge generational divides and improve understanding between generations, the frequency and caliber of intergenerational exchange sessions are also tracked. The Forum of Expertise uses visual documentation in addition to numerical data to document the qualitative impact of its programs. Videos and pictures of different activities offer a genuine portrayal of social interactions and participant engagement. These visual records are useful resources for evaluating the initiative's emotional and psychological effects in addition to being promotional materials. Program facilitators' observational evaluations also help to reveal the breadth of social ties forged via the Forum.

Frequent feedback loops guarantee that measurements guide the initiative's continued development. Through surveys and casual conversations, participants are encouraged to share their experiences, providing valuable insights into the program's most beneficial aspects and areas for improvement. Activities can be modified thanks to these feedback systems, guaranteeing that they continue to be impactful and relevant for all participants. For long-term planning, adaptation, and program improvement, the gathered information and observations are essential. Through the examination of volunteerism, engagement trends, and intergenerational interactions, the Forum of Expertise adapts to its members' needs over time. This evidence-based strategy guarantees that the program will continue to be successful in encouraging active aging in the community, lowering isolation, and promoting social inclusion.

13.2.3 Implementation

13.2.3.1 Principal Parties and Partners

Strategic partnerships that bring together a variety of stakeholders enable the Forum of Expertise to be implemented successfully. Each stakeholder contributes resources, expertise, and support to further the initiative's objectives. This initiative is spearheaded by the Ramallah Municipality, which promotes a cooperative framework that puts the welfare of the elderly first while guaranteeing sustainability and effectiveness.

The Forum of Expertise and private medical firms have a significant partnership that creates a novel approach to senior healthcare. These partnerships provide screenings, medical consultations, and customized wellness programs with an emphasis on the early detection, prevention, and management of age-related health conditions. The collaboration guarantees that participants receive prompt medical attention by incorporating healthcare support into the program, lowering long-term health risks and improving overall quality of life. By utilizing the experience of private healthcare facilities, this strategy also maximizes resources while remaining economical. Transparency is the foundation of the partnership, guaranteeing that business interests won't override the main goal of advancing senior health and wellbeing.

In addition to providing healthcare, the Forum actively works with youth clubs, neighborhood non-profits, and charitable organizations to build an inclusive environment where senior citizens can take part in worthwhile social and community-focused activities. Through these collaborations, senior citizens can participate in community service projects, skill-sharing seminars, and mentoring programs, reaffirming their value as contributing members of society. The program encourages intergenerational relationships by collaborating with youth organizations, enabling the elderly to stay involved and active while also enabling younger generations to gain from the knowledge and experiences of more senior community members.

The initiative's reach is further strengthened by the partnership with institutions and organizations located in the city. Social service providers, educational institutions, and cultural centers all make contributions by providing areas, learning materials, and organized programs that support the Forum's goals. In addition to receiving care and support, this integrated approach guarantees that senior citizens actively participate in forming their communities. The Forum of Expertise creates an impact-maximizing, balanced model by bringing together the public sector, private businesses, non-profits, and community organizations. Through addressing their healthcare needs, promoting social inclusion, and reaffirming their crucial role in the city's social fabric, the combined efforts of these key players guarantee that elderly residents of Ramallah receive comprehensive support.

13.2.3.2 Resources for Implementation

A coordinated combination of human, financial, and physical resources from various industries is necessary for the Forum of Expertise to be realized successfully, guaranteeing efficiency and sustainability. The initiative's physical infrastructure, a special building in Ramallah's historic old city, is at its heart. In order to create a welcoming and accessible environment for senior participants, this space was first renovated thanks to a generous donation from the Global Community. In addition to maintaining the building's historical integrity, this donation created a long-term center for active aging initiatives and intergenerational interaction.

The Ramallah Municipality provides the Forum's operating funds, demonstrating the city's dedication to promoting inclusive urban development and social cohesion. This municipal support, which covers necessary expenses like utilities, program coordination, and facility maintenance, guarantees the initiative's long-term sustainability. To optimize impact, the initiative strategically integrates community-driven resources in addition to financial support.

The Forum's dependence on senior volunteers, who actively participate in planning and directing events, is one of its distinguishing characteristics. These volunteers enhance the project and turn it into a dynamic, self-sustaining model by contributing a multitude of knowledge, abilities, and life experiences. Among their contributions are peer support, workshop instruction, mentoring, and organizing leisure activities. A sustainable strategy that lessens reliance on outside funding while boosting the initiative's legitimacy and relevance is demonstrated by this integration of senior knowledge and leadership.

The Forum receives in-kind donations from nearby companies, nonprofits, and community members in addition to municipal funding and volunteer participation. These donations include medical supplies, equipment, educational materials, and logistical assistance—all of which are critical for sustaining and growing program offerings. The program guarantees operational effectiveness, financial sustainability, and widespread community involvement by utilizing a wide range of resources. The initiative's sustained success is supported by a robust and flexible framework that is produced by this multi-layered resource strategy, which combines community involvement, private sector contributions, and public investment. By taking a comprehensive approach, the Forum of Expertise not only flourishes as a hub for senior empowerment but also acts as a template for resource-efficient, community-driven urban initiatives.

13.2.3.3 Tools and Methods Applied

To guarantee its success in promoting senior engagement and well-being, the Forum of Expertise combines structured methodologies with flexible strategies. Increased participation rates, greater volunteer involvement, and a developing culture of knowledge sharing among senior members are all concrete indicators of the initiative's impact. Self-reported increases in well-being, which are recorded through participant stories, direct feedback, and qualitative evaluations, serve as additional evidence of these programs' efficacy. Seniors who participate in the Forum often report feeling more fulfilled personally, connected to others, and having a renewed sense of purpose.

In order to accomplish these goals, the Forum consistently broadens the scope of its operations to make sure they continue to be pertinent and appealing to a wide range of senior citizens. Programs are made to accommodate a range of interests, such as intellectual stimulation through workshops and discussion groups, physical well-being activities like yoga and light exercise, and creative engagement through storytelling, music, and the arts. Participants' interest is maintained and stagnation is avoided with this dynamic and changing approach. A significant part of the Forum's methodology involves interactions between generations. The program actively encourages senior citizens and younger community members to work together, which fosters social cohesion and mutual learning. This is achieved through structured mentorship programs, skill-sharing workshops, and volunteer-led initiatives that bridge generational gaps. By promoting these interactions, the Forum not only boosts senior citizens' sense of self-worth but also fortifies the social cohesion of the larger community.

The Forum conducts frequent assessments and evaluations to guarantee ongoing advancement toward its objectives. Participant feedback surveys, volunteer engagement monitoring, and attendance records are examples of data collection methods. These resources offer insightful information about the efficacy of programs, assisting planners in improving current endeavors and launching fresh projects in response to changing demands. In order to obtain detailed viewpoints on the difficulties and experiences encountered by senior participants, facilitators also occasionally hold focus groups and one-on-one interviews. The iterative nature of these assessments ensures that the Forum remains flexible and responsive. Adjustments are made in real-time, based on emerging trends and participant input, allowing for continuous improvement and meaningful progress toward global objectives such as active aging, social inclusion, and intergenerational solidarity. Through these well-structured tools and methodologies, the Forum of Expertise serves as a replicable model for community-driven elderly engagement programs.

13.3 Obstacles and Solutions

The impact of the COVID-19 pandemic and the low level of digital literacy among the elderly were the two most pressing issues facing the Forum of Expertise during its implementation. Senior participants were at a higher risk of social isolation as a result of the pandemic's severe disruption of in-person activities. Remote engagement was also challenging because many senior members lacked the skills needed to use digital platforms. The initiative's primary goal of encouraging social interaction and active aging was in danger of being compromised by these issues.

By incorporating digital literacy training into its programming, the Forum took a proactive and ingenious approach to resolving these problems. Elderly participants received one-on-one assistance from young volunteers who served as mentors, teaching them how to use social media, video conferencing apps like Zoom, and smartphones. In addition to bridging the technological divide, this mentorship approach improved ties between generations and reaffirmed the Forum's dedication to social inclusion.

The initiative launched creative online programs to keep senior members involved during the lockdown. In order to keep participants connected to their heritage and artistic expressions, two noteworthy programs—Nasma wa Nastamteh and Raseef 18—offered virtual cultural experiences. Raseef 18 presented discussions with 18 creative writers, encouraging intellectual engagement among the senior audience, while Nasma wa Nastamteh included interviews with 35 Palestinian musicians, providing insights into their artistic journeys. These shows reduced feelings of loneliness and alienation by offering an essential source of inspiration, education, and entertainment. The Forum used well-known social media and messaging apps to keep in touch and provide support. Specialized Telegram, WhatsApp, and Facebook groups were established as interactive forums where senior citizens could converse, exchange stories, and get updates. Even in situations where in-person meetings were not feasible, these online communities offered an extra degree of assistance by enabling participants to remain informed and connected.

The Forum of Expertise was able to overcome obstacles brought about by the pandemic and technological inaccessibility by using these flexible tactics. The initiative's sustained success was guaranteed by its capacity to quickly adopt digital solutions, encourage intergenerational cooperation, and develop novel approaches to engagement. In addition to resolving immediate issues, these solutions gave senior citizens long-lasting digital skills that will enable them to stay involved and connected in a world that is becoming more and more digital.

13.3.1 Achieved Outcome

With important results that support its primary goal of encouraging engagement, well-being, and social inclusion, the Forum of Expertise's Active Aging initiative has significantly improved the lives of senior citizens in Ramallah. The notable rise in senior community involvement and volunteerism has been one of the main successes. The program has effectively established a forum where senior citizens feel encouraged to give back to the community by sharing their life experiences and expertise in addition to volunteering. Their sense of purpose and interpersonal connections have been strengthened by this active participation, which has enhanced their overall well-being.

The empowerment of seniors, many of whom have discovered a new purpose in mentoring others and sharing their knowledge, is another significant accomplishment of the Forum. Elderly people have been able to receive community recognition as a result of this process, which has confirmed their value and worth. Once disregarded, seniors are now revered sources of knowledge that promote intergenerational learning and dispel myths about aging. Additionally, by organizing physical activities like yoga and light exercise, the initiative has been instrumental in encouraging seniors to lead active lifestyles. Elderly participants' physical health has improved as a result of these programs, which has enhanced their mobility, mental health, and general quality of life. For Palestinian seniors living in urban areas, who frequently encounter obstacles in obtaining healthcare and fitness resources, these physical activities are especially important.

The Forum has made it a priority to preserve and share the stories of its senior members in addition to its health-related initiatives. Its first book, which chronicled the experiences of 25 participants, was released this year. This book is a tribute to the tenacity and contributions of Palestinian seniors as well as a historical account of their lives and experiences. Elderly people now have a voice thanks to the Forum's documentation of these personal stories, which has promoted their dignity and recognition in the larger community. Through these initiatives, the Forum of Expertise has been able to accomplish significant results that have improved the quality of life for senior citizens and made society more cohesive and inclusive.

13.4 Lessons Learned

The Forum of Expertise has experienced a journey of transformation, filled with important lessons learned, especially from the difficulties presented by the COVID-19 pandemic. The crisis made clear how crucial flexibility and foresight are to maintaining community involvement. The need for technology integration to sustain engagement and connection, even during lockdowns, was one of the most important lessons learned. The initiative's choice to provide senior citizens with technology training made sure they stayed engaged and connected, enabling them to take part in online workshops, meetings, and activities. This change in strategy not only closed the digital divide but also gave senior members more self-assurance and independence, highlighting the value of digital literacy for seniors in a world that is becoming more interconnected by the day.

The pandemic highlighted the value of local cooperation and volunteerism in addition to technological advancements. The Forum was able to assist elderly people who were more likely to be isolated during lockdowns by organizing volunteers. People from all walks of life came together to ensure the well-being of the elderly, strengthening the community's resilience and fostering a sense of solidarity. The idea that crises can spur innovation was reaffirmed by this experience, and the Forum showed that even during challenging times, innovative solutions—like creating online documentaries and conducting virtual interviews—could encourage participation and preserve a feeling of community.

Furthermore, the Forum of Expertise has developed into a template that has served as an inspiration to nearby municipalities in Palestine and abroad. Similar models have been used by the municipalities of Al Bireh, Selfeet, and Madaba, Jordan, to set up their own facilities devoted to community involvement and active aging. The initiative's success and the universal applicability of its methodology are demonstrated by this regional expansion. The significance of involving senior citizens in community projects at an early age is a crucial lesson to be learned from this. The Forum has strengthened intergenerational solidarity and empowered seniors by acknowledging their worth and utilizing their knowledge.

The Forum's success story offers other cities a model to follow, with useful advice on how to design inclusive environments that promote resilience, wellbeing, and deep connections. Its strategy emphasizes how older generations can benefit society by enhancing communities with their knowledge, anecdotes, and abilities. The Forum's influence will surely be felt globally as it shares its expertise through workshops, consultations, and cooperative projects, encouraging cities to realize the full potential of their senior citizens. Building resilient, connected communities is both feasible and necessary for a sustainable future, as the Forum of Expertise exemplifies through its dedication to inclusivity and intergenerational dialogue.

Related SDGs: Goals 3 and 10

Acknowledgements We explicitly thank the project owners based in Ramallah, Palestine, for their participation and the provision of materials for the 6th Guangzhou International Award for Urban Innovation. We also congratulate them on their achievements in the 2023 cycle of awards.

The above images are provided to us by the project owners for their submission.

Chapter 14
Fomenting Entrepreneurship and Formal Employment Processing Through Programa Tô Legal in São Paulo, Brazil

Abstract The "Tô Legal" electronic system was designed as a platform to empower individuals seeking formal employment opportunities despite lacking the typical qualifications, education, or experience usually required. Many of these individuals produce their own goods and services, often enhancing both their skills and products over time, which in turn allows them to transition from informal public spaces to formalized work, either within companies or as micro-entrepreneurs. Through this system, a new type of document—the Authorization Ordinance for Commerce and Service Provision (PCS)—was introduced. This document, which allows citizens to operate in public spaces for a maximum of ninety days, is easily obtained through automated application analysis. As long as the dates and times of these authorizations do not conflict, there is no limit on the number of PCS authorizations a person may possess. Eleven different permits and three different kinds of authorizations are among the fourteen documents that the system generates; they are all pertaining to the control of business and service delivery on sidewalks and public areas. This case study serves as an example of how digital platforms can help marginalized communities in São Paulo achieve economic inclusion, legal recognition, and dignity. The initiative is related to SDGs 8, 9, 10, 11, 12, 16, and 17.

Keywords Tô legal · Electronic system · Public spaces · Micro-entrepreneurship · Digital inclusion · Informal work · Regulatory innovation

14.1 Introduction to São Paulo, Brazil

The world-renowned metropolis of São Paulo, the biggest city in Brazil and the Southern Hemisphere, is renowned for its vibrant urban environment, diverse culture, and robust economy. As the financial and commercial center of Brazil, São Paulo is the most populous city in the country, home to about 12.3 million people. The city's highly urbanized landscape is reflected in its large area of 17,000 km^2 and high population density of 24,821 people per square kilometer.

A. Cheshmehzangi et al., *People-Centered Innovation*, Urban Sustainability,
https://doi.org/10.1007/978-981-96-5230-3_14

São Paulo's notable 11% population growth rate highlights the city's continuous urbanization and economic prospects, which draw migrants from all over Brazil and beyond. A vibrant social fabric is created by the city's multicultural population, which is heavily influenced by Asian, African, and European communities. Notwithstanding its enormous size and expansion, São Paulo still has to deal with urban issues like a lack of housing, heavy traffic, and socioeconomic inequality. São Paulo is one of the wealthiest cities in Latin America, with a high GDP per capita of $60,750. As Brazil's financial hub, the city is home to the main offices of important international firms, top banks, and the B3, the biggest stock exchange in Latin America. With robust sectors in manufacturing, services, technology, finance, and commerce, the city's economy is diverse. São Paulo also plays a significant role in global trade and investment, drawing companies and entrepreneurs from all over the world.

São Paulo faces severe social inequality despite its economic prosperity, as evidenced by its GINI Index of 47.1 (as of 2021). This measure draws attention to the contrast between low-income populations living in informal settlements, or "favelas," and the affluent neighborhoods of places like Jardins and Vila Olímpia. Even though the city has top-notch healthcare, educational, and infrastructure facilities, access to these services is still uneven, which exacerbates socioeconomic gaps that have existed for a long time. Highways, bus systems, and the nation's largest metro system are all part of São Paulo's vast but overworked transportation network, which is the most developed urban center in Brazil. However, with daily commutes frequently lasting several hours, traffic congestion continues to be a persistent problem. To solve these problems, the city keeps spending money on urban mobility and public transportation. Rapid urbanization has also raised issues with environmental sustainability, such as air pollution, water scarcity, and deforestation in nearby areas.

Beyond its economic importance, São Paulo is a cultural powerhouse that is home to well-known organizations like the São Paulo Biennial, the Ibirapuera Park, and the São Paulo Museum of Art (MASP). The city draws tourists from all over the world because of its thriving nightlife, dining, and music scenes. With its blend of modern urban life and historical legacy, São Paulo is a city of contrasts, where social issues and economic opportunity coexist in one of the world's most vibrant metropolitan areas.

14.2 Background and Origins

Although São Paulo has long been a city of opportunity, historically it was difficult and constrained for informal economic activities to access legal channels. Before 2019, street vendors and service providers could only lawfully operate in public areas with a Use Permit Term (TPU), which was a challenging permit to obtain because of its limited availability and bureaucratic nature. Permits were only issued for specific locations chosen by the local deputy prefect, and the procedure required a public convocation or application. Unemployed people, young people, and senior employees without access to formal employment were frequently marginalized by this system.

The municipal government introduced Tô Legal, an online platform created to expedite the process of obtaining business permits and authorizations for commerce and service provision in public spaces, after realizing the need for a more accessible and inclusive approach. The system makes it easier for prospective business owners, including underrepresented populations like the elderly, people with disabilities, former prisoners, and veterans of war, to legalize their operations. Tô Legal greatly lowers bureaucratic barriers and promotes local entrepreneurship by issuing Authorization Ordinances that permit holders to operate on sidewalks and in designated public areas.

By providing 11 distinct TPU types, each subject to distinct laws, the Tô Legal system democratizes access to business opportunities. With the help of an interactive Geographic Information System (GIS) map that shows approved business operations throughout the city, the platform gives applicants the freedom to select the locations and operating hours that work best for them. When choosing a location for their business, entrepreneurs can use this visualization tool to test potential sites and determine their viability before settling on one. To ensure that their operations are approved by the law, people must first register as Individual Micro-Entrepreneurs (MEIs) or with the Municipal Treasury Office before they can access Tô Legal. After registering, users can choose where and how long they want to work; permits are issued for up to 90 days. Additionally, the system increases transparency by providing users, municipal office administrators, and regulatory inspection staff with access to permit-related data.

In addition to streamlining the licensing procedure, Tô Legal has emerged as a vital instrument for advancing São Paulo's economic inclusion. A degree of flexibility that is frequently absent from traditional employment models is offered by the system, which enables people to personalize their work schedules by choosing particular days and times. Additionally, by digitizing records and enabling real-time monitoring of economic activities, it improves the relationship between entrepreneurs and local authorities. Additionally, the platform includes a payment control grid that lets administrators see georeferenced permit locations throughout the city and monitor revenue collection. This feature guarantees that São Paulo's informal economy functions within a transparent and organized framework while also improving financial accountability.

The Tô Legal system is a prime example of how digital transformation can promote social equity, urban resilience, and economic empowerment as São Paulo develops. The platform has strengthened the city's economic fabric and addressed long-standing issues of bureaucracy and exclusion by making it easier for thousands of informal workers to obtain business permits. Projects like Tô Legal are a model for other international cities looking to promote inclusive urban economies in a city as vibrant as São Paulo.

14.2.1 Key Innovative Aspects

São Paulo, a city known for its thriving economy and spirit of entrepreneurship, has long struggled to incorporate its sizable informal sector into a formal regulatory framework. The Tô Legal system, which uses technology-driven solutions to expedite permit issuance and improve economic inclusion, represents a revolutionary change in urban governance. The way street vendors and service providers get legal permission to operate in public areas has been completely transformed by Tô Legal through the use of Geographic Information Systems (GIS), real-time data analysis, and seamless digital integration.

Tô Legal's reliance on technology, especially database integration and GIS mapping, is one of its key innovations. Fifteen distinct databases are connected to the system, guaranteeing transparent and effective processing of all applications. An interactive digital map that shows all of São Paulo's approved business zones allows entrepreneurs to choose the place they want to work. Based on current market conditions, competition, and customer traffic, this feature enables users to make well-informed decisions about their commercial activities.

Tô Legal does all of its business online, in contrast to the old, cumbersome TPU procedure that required in-person applications and lengthy wait times. From the comfort of their homes, applicants can submit requests, get prompt analysis, and print their Permission Certificate for Services (PCS) after it has been approved and paid for. This makes the system much more effective and user-friendly by doing away with the necessity for in-person trips to government offices.

Real-time permit processing is one of the biggest innovations created by Tô Legal. In the past, vendors had to deal with drawn-out bureaucratic processes, and approvals frequently took months. Instantaneous analysis of applications has improved access to legal permits and decreased wait times. For informal workers, this innovation increases predictability, enabling them to confidently plan their businesses. Moreover, the platform promotes government transparency by centralizing data collection and financial transactions. All issued permits, payment records, and location-based authorizations are stored within the system, minimizing corruption risks and ensuring that permit allocation follows a fair and systematic process. The digital payment system also prevents informal financial exchanges, further strengthening regulatory compliance.

In a city where bureaucratic procedures have traditionally necessitated in-person visits to government offices, Tô Legal has revolutionized the accessibility of public services. The municipal administration of São Paulo has shifted to e-governance and drastically cut red tape by digitizing permit issuance. Through automated reports, this change not only improves user convenience but also strengthens budgetary control, allowing policymakers to better monitor revenue collection and distribute resources.

Tô Legal is notable for its dedication to economic empowerment in addition to its technological innovations. With the platform's unparalleled flexibility, users can choose their own work schedules, picking the days of the week, time slots, and operating hours that work best for them. Before making long-term commitments, entrepreneurs can also test out various locations and business models using the system's trial mechanism. This lowers business risk, particularly for new business owners, and promotes innovation and economic expansion in the community. Furthermore, Tô Legal's role in promoting inclusive urban economies is highlighted by the inclusion of socially vulnerable groups, including the elderly, people with disabilities, former prisoners, and veterans of war. The platform acts as an example of social innovation in local government by offering organized and easily accessible opportunities for self-employment.

Tô Legal has raised the bar for urban economic governance by using technology to streamline bureaucratic procedures, improve transparency, and empower small business owners. São Paulo is now a leader in smart city innovation thanks to its integration with GIS, real-time data processing, and digital payment systems, which has revolutionized the regulation of public spaces. In addition to helping entrepreneurs, the project's focus on adaptability, inclusivity, and efficiency fortifies municipal oversight and opens the door to a more vibrant and sustainable urban economy.

14.2.2 Desired Outcome and Measurement

14.2.2.1 Desired Outcome

The Tô Legal project was created with a specific set of goals in mind: to improve governance transparency, encourage economic inclusion, and expedite administrative procedures. Fundamentally, the system aims to make it easier and faster to issue licenses and authorizations for business and service delivery in São Paulo's public spaces, making the city a more welcoming and equal place for business owners. Tô Legal empowers people who have trouble breaking into the formal labor market, such as the unemployed, young people, and senior employees, by lowering bureaucratic obstacles, making it easier for them to start legally recognized businesses.

By offering a legal framework for informal workers and 14 different types of permits that accommodate a variety of economic activities, the system aims to promote entrepreneurship. The platform's digitization of the permit issuance process removes needless delays, enabling users to quickly obtain work authorization and launch their businesses without undue red tape. More financial independence and job creation result from this, especially for underserved populations. Tô Legal wants to improve urban management and governance in addition to economic empowerment. The system maximizes the use of public space while upholding order and compliance by ensuring that economic activities are carried out in authorized areas through its GIS-based location monitoring. Additionally, the digital tracking of permits helps with urban planning by giving policymakers useful information to control the distribution of commerce and gauge the demand for particular services throughout São Paulo.

Improving financial transparency in order to fight corruption is another important goal. By removing unofficial transactions, the platform's centralized payment control system guarantees that all permit fees are collected in a controlled and traceable manner. As a result, there are fewer opportunities for bribery and favoritism, promoting a just and equitable system in which permits are issued according to precise, uniform standards rather than personal ties. Increasing operational efficiency in municipal administration is another goal of the project. Public officials' workload is lessened by Tô Legal's digitization and automation of permit issuance, freeing them up to concentrate on strategic planning and oversight rather than administrative bottlenecks. This guarantees that the system can manage large application volumes without experiencing any delays, in addition to improving service delivery.

The success of Tô Legal is measured through several key performance indicators (KPIs), including:

- Reduction in permit processing time: Tracking the time it takes from application submission to approval.
- Increase in the number of legalized entrepreneurs: Assessing how many previously informal workers have entered the formal economy.
- Revenue collection through centralized payments: Evaluating financial transparency and proper allocation of municipal funds.
- GIS-based location data: Measuring the effectiveness of location monitoring in managing public space use.
- Public satisfaction and accessibility: Analyzing feedback from users to ensure ease of access and efficiency.

By achieving these outcomes, Tô Legal not only enhances São Paulo's economic landscape but also strengthens institutional integrity, promotes sustainable urban growth, and fosters an inclusive entrepreneurial environment for all residents.

14.2.2.2 Measurement

The Tô Legal project is designed not only to facilitate permit issuance but also to ensure that its impact is measurable and transparent. To achieve this, a comprehensive monitoring system has been implemented, enabling the continuous evaluation of performance and effectiveness.

To guarantee compliance, effectiveness, and equity, the managing team of Tô Legal monitors the system every day and keeps track of 14 distinct kinds of permits and authorizations. Municipal administrators have access to a dashboard that gives them a clear picture of the system's performance and includes real-time graphics and revenue data. Authorities can track the initiative's progress, spot bottlenecks, and take proactive measures to resolve possible problems before they worsen thanks to this centralized data collection. The measurement system's impersonality and objectivity are important characteristics. Tô Legal ensures that permit allocation is free from discrimination by not collecting age, race, or gender data, in contrast to traditional bureaucratic processes that might be biased. Rather, the system collects information on location, activity type, and equipment, which aids in the analysis of São Paulo's economic geography by municipal authorities. In order for the city to evaluate the economic potential of various locations and distribute resources appropriately, this information is essential for urban planning and policy development.

Revenue collection via the centralized payment system is another important indicator. Administrators can guarantee that all permit fees are appropriately collected, minimizing financial irregularities and improving municipal budget planning, by monitoring payments in real time. Additionally, by providing information about which business sectors are expanding and where more support might be required, this data aids in forecasting economic trends. Additionally, the georeferenced permit data offers a graphic depiction of the city's economic activity. Policymakers can identify high-demand areas, underutilized spaces, and possible over-concentrations of specific business types by using the system to map out authorized commerce locations. This facilitates strategic decision-making by guaranteeing that urban policies correspond with São Paulo's real economic dynamics.

The success of Tô Legal is measured through various quantitative and qualitative indicators, including:

- Processing time per permit: Evaluating how quickly applications are approved.
- Number of issued permits: Tracking legal business growth over time.
- Revenue collection efficiency: Ensuring all payments are properly accounted for.
- Location-based economic data: Assessing commerce distribution and potential gaps.
- User satisfaction and accessibility: Gathering feedback from entrepreneurs on system usability.

By continuously monitoring, analyzing, and refining its operations, Tô Legal ensures that São Paulo's public spaces are efficiently managed, economically vibrant, and accessible for all aspiring entrepreneurs.

14.2.3 Implementation

14.2.3.1 Principal Parties and Partners

The Tô Legal project's execution in São Paulo depends on a network of important stakeholders and partners cooperating to improve service delivery, expedite administrative procedures, and assist regional companies. In order to ensure effective data processing and the ongoing development of the city's administrative database, the Municipal Treasury's cooperation with other municipal offices is at the heart of this endeavor. These collaborations aid in real-time information updating, inconsistency correction, and the eventual removal of administrative roadblocks that frequently impede the provision of public services. Tô Legal can expedite approvals and provide a more transparent and easily accessible system for informal and small-scale entrepreneurs by utilizing these governmental partnerships.

In addition to government cooperation, the project gains from a strategic partnership with the Commerce and Trade Association of São Paulo. Understanding Tô Legal's capacity to promote economic development, the association has suggested a collaboration to raise knowledge and accessibility of the Program for the City of São Paulo (PCS) and other Tô Legal-issued official documents. The association sees itself as a center where small business owners can get advice on formalization, legal compliance, and business development by serving as a liaison between the initiative and local entrepreneurs. The support system for unofficial vendors and new businesses looking for legitimacy within the city's regulatory framework is strengthened by this partnership, which also raises awareness of Tô Legal.

These partnerships work together to build a vibrant ecosystem that advances the project's overarching objectives of streamlining administrative procedures, enhancing regulatory compliance, and cultivating an atmosphere that supports the growth of small enterprises. The participation of trade associations and municipal authorities guarantees the initiative's technical efficiency and broad accessibility, which ultimately contributes to São Paulo's more formalized and inclusive economic landscape.

14.2.3.2 Resources for Implementation

The financial resources provided by the District's Secretariat's annual budget have enabled the Tô Legal project to be implemented in São Paulo. This funding covers essential elements like system maintenance, error correction, and the integration of new functionalities, ensuring the initiative's stability and continuity. The project gains a stable financial framework that permits continuous enhancements and long-term viability by obtaining a specific budget within the municipal structure.

Formal support for Tô Legal is provided by a contract with the District Secretariat that ensures the platform will be maintained. In order to maintain the system's effectiveness, timeliness, and responsiveness to user demands, this contract specifies the requirements for routine upgrades. The project's reach and efficacy for small businesses and informal vendors looking for legal recognition can be increased by adding more tools and documents to its scope as new regulatory requirements arise or as demand for additional services increases.

Sustaining user confidence and accessibility requires a dedication to system improvement and troubleshooting. The initiative guarantees a smooth digital experience for entrepreneurs navigating the formalization process by dedicating resources to resolving technical issues and improving performance. Technological developments are also made possible by the organized financial support, which keeps Tô Legal in step with São Paulo's changing administrative and economic environments.

Tô Legal keeps growing as a dependable and expandable business formalization solution in the city thanks to these well-allocated resources. Its incorporation into the budget of the District's Secretariat and the contractual structure that underpins its operations demonstrate a calculated approach to maintaining and growing the project, which eventually strengthens its function in cutting red tape and promoting economic inclusion.

14.2.3.3 Tools and Methods Applied

The Tô Legal project in São Paulo has used a variety of tools and techniques to enhance governance and ease business formalization since its inception on July 1, 2019. The program functions as an online platform that reduces bureaucracy and increases transparency in municipal administration by expediting the issuance of permits and authorization ordinances. Tô Legal makes it possible for street vendors, small business owners, and service providers to swiftly and effectively acquire the required legal documents by utilizing digital tools. For informal workers, this strategy has greatly improved accessibility, enabling them to enter the formal economy with little administrative hassle.

Between its inception to October 14, 2023, the platform has issued 55,829 Authorization Ordinances for Commerce and Service Provision, along with 9565 Use Permit Terms. These figures demonstrate how Tô Legal has been successful in formalizing businesses throughout São Paulo, giving more than 19,000 ordinance and permit holders legal recognition. With over R$72.5 million (roughly US$14 million) from various municipal charges, such as nine different types of TPUs and three types of Authorization Permits, as well as fees associated with street, farmer's, and produce markets, the system's effectiveness has also helped the municipality collect revenue.

Tô Legal's approach is in line with more general objectives for economic development and governance. By giving entrepreneurs—especially those in the unorganized sector—a gateway, the program is steadfastly dedicated to promoting social and economic development. The project encourages small businesses to comply with municipal requirements by streamlining regulatory processes, which gives them legitimacy and opens up new opportunities. By doing this, Tô Legal contributes to the development of a more welcoming business climate where microbusiness owners can prosper without worrying about excessive red tape or legal obstacles.

The project's contribution to improving governance and thwarting corruption is another essential component of its execution. Tô Legal enhances accountability and reduces the likelihood of illegal activity by digitizing permit issuance and decreasing in-person interactions in administrative procedures. The system makes sure that every transaction is documented, resulting in an open and verifiable record of payments and authorizations. By providing a fair and uniform permitting process, this approach not only improves municipal oversight but also fosters confidence among business owners.

Tô Legal also acts as a responsive instrument in emergency situations. The platform can adjust to economic and social challenges by providing a structured yet adaptable digital solution, guaranteeing that entrepreneurs receive support even in unpredictable situations. The initiative continues to be a vital tool for São Paulo's business community, whether it is through expediting permit renewals or adapting regulations to suit changing market conditions. Tô Legal continues to revolutionize the urban and business landscape of São Paulo by integrating digital tools, transparent processes, and economic facilitation. The initiative serves as a model for inclusive development and efficient governance by streamlining formalization, decreasing corruption, and promoting economic resilience.

14.3 Obstacles and Solutions

There was an initial period of adaptation for both permit applicants and public office employees as a result of the Tô Legal project's implementation in São Paulo, which brought about significant changes to the way application analyses and document issuance procedures were carried out. Changes in administrative procedures and user interactions with the municipality's regulatory framework were necessary for the shift to a new digital system. Although the project's goal was to streamline and simplify bureaucratic procedures, there were obstacles to overcome in order to successfully implement the transition from conventional techniques to a fully digital system.

The project's managing team overcame these challenges by putting in place a number of calculated fixes intended to promote adaptability and enhance user experience. Setting up internal training sessions for staff members in public offices was one of the main actions done. These training sessions gave employees the skills they needed to process applications accurately, navigate the new system quickly, and help citizens more successfully. The program made sure that municipal staff could confidently handle the new digital processes while preserving service quality by funding civil servant capacity-building.

Tô Legal also implemented open office hours so that citizens could get direct assistance and clarification on the application process, acknowledging that many small business owners and entrepreneurs were not familiar with the new system. Through these in-person consultations, applicants were able to get answers to their questions, get advice on the necessary paperwork, and more easily navigate the new digital platform. Users who were switching from paper-based applications to an online system felt more confident and less confused as a result of this proactive engagement.

The project team created thorough user manuals for both citizens and municipal employees in addition to providing direct assistance. These manuals are a permanent resource for anyone who needs help with the system and can be downloaded from the internet and intranet portals. These guides assist users in independently completing their applications and comprehending the regulatory requirements by offering concise, detailed instructions. This minimizes the need for repeated questions and guarantees more seamless platform interactions.

Tô Legal was able to successfully address the implementation challenges by implementing these focused solutions. All parties involved experienced a more seamless transition thanks to the combination of training initiatives, public outreach campaigns, and readily available informational materials. The project's emphasis on support and education not only made it easier to adopt a more effective digital system, but it also reaffirmed São Paulo's dedication to enhancing governance and service accessibility.

14.3.1 Achieved Outcome

In São Paulo, the Tô Legal project has produced noteworthy results, revolutionizing bureaucratic processes and improving accessibility for business owners and local government officials. All citizens are impacted by this municipal policy, which streamlines business formalization and boosts government effectiveness. Entrepreneurs no longer need to physically visit sub-prefectures to apply for documents, update their registry, or pay municipal fees thanks to the initiative's digitization of application processes. Now offered online, these services have lessened administrative demands, saving entrepreneurs' time and money while fostering a more open and approachable regulatory framework.

The system has greatly accelerated inspection processes by giving civil servants instant access to permit and authorization data. Tô Legal has increased regulatory oversight and increased the effectiveness of enforcement by making sure that officials can track authorizations in real time and verify compliance. Public offices can now function more precisely and quickly thanks to the ability to access current records without depending on paper-based documentation, which has also decreased errors and processing delays.

The initiative's optimization of municipal revenue collection is another important result. Tô Legal has made the financial system more transparent and effective by automating payment procedures and sending money directly to the Municipal Treasury. By offering a transparent and verifiable record of all payments, the digital platform improves fiscal accountability and lowers the risks connected with manual transactions. As a result, São Paulo's financial management has improved, guaranteeing that municipal funds are gathered on time and distributed to public services efficiently.

All things considered, the project has brought São Paulo's business regulation practices up to date and made the process more effective, transparent, and user-friendly for all parties involved. Business formalization is made easier for entrepreneurs by a streamlined, entirely digital process, and revenue management and administrative efficiency are enhanced for local government agencies. Tô Legal exemplifies efficient governance by cutting red tape, speeding up inspections, and simplifying financial transactions, thereby reaffirming São Paulo's dedication to promoting economic expansion and regulatory innovation.

14.4 Lessons Learned

The Tô Legal project in São Paulo has provided insightful lessons on economic resilience, transparency, and governance while showcasing the potential of digital transformation in public administration. The system, which was put into place with a solid foundation in efficiency and innovation, has improved revenue collection, decreased bureaucracy, and expedited business formalization. Its effects were especially noticeable during the pandemic years, when the city encountered previously unheard-of financial difficulties. Mayor Bruno Covas oversaw the signing of decrees lowering municipal fees for the PCS and other permits, guaranteeing that small business owners and unorganized laborers could carry on with their operations in spite of the crisis.

The livelihoods of thousands of workers were protected as the PCS was promptly reissued, even after brief suspensions brought on by lockdowns. For the citizens of São Paulo, the Tô Legal system has proven to be a quick and effective way to expand economic opportunities and implement government policies. The significance of process simplification, which benefits both citizens and civil servants, is one of the project's main lessons. The initiative has improved municipal oversight and empowered entrepreneurs by making regulatory processes more transparent and accessible.

Beyond efficiency, Tô Legal has improved governance for the municipality by implementing centralized revenue collection, real-time debtor mapping, and thorough budget reporting. By eliminating in-person interactions during the permitting process and guaranteeing that every transaction is digitally recorded and traceable, it has also been instrumental in the fight against corruption. These systemic changes demonstrate how digital tools can improve public institutions' accountability and credibility.

The diligence of Tô Legal's small but incredibly productive administrative staff—four civil servants operating in a dedicated, all-female division—is responsible for the company's success. This team has issued over 40,000 documents and is still managing and updating the system with accuracy and agility despite having few resources. Their dedication to ongoing development is indicative of a progressive approach to public service, where creativity is fueled by flexibility and real-world demands. Through constant improvement of digital tools and user experience, the team has made sure that Tô Legal continues to be a model of accessibility and efficiency.

In the future, other cities looking to improve economic inclusion and modernize public administration can benefit greatly from the project's five years of experience. The group is still dedicated to increasing the system's functionality, producing additional documents, and looking into fresh approaches to enhancing public services.

Their work serves as an example of how strategic digital transformation can produce long-term advantages for urban resilience, economic growth, and governance.

Related SDGs: Goals 8, 9, 10, 11, 12, 16, and 17.

Acknowledgements We explicitly thank the project owners based in São Paulo, Brazil, for their participation and the provision of materials for the 6th Guangzhou International Award for Urban Innovation. We also congratulate them on their achievements in the 2023 cycle of awards.

Autorização para Comércio e Serviços
Valet
Mesas / Cadeiras / Toldos
Compartilhamento de Bicicleta
Comida de Rua
Flores
Banca de Flores
BANCA DE JORNAL
Banca de Jornais e Revistas
Ambulante
Compartilhamento de Patinetes
Porta a Porta
Feira Livre
MERCADO
Mercado
SACOLÃO
Sacolão

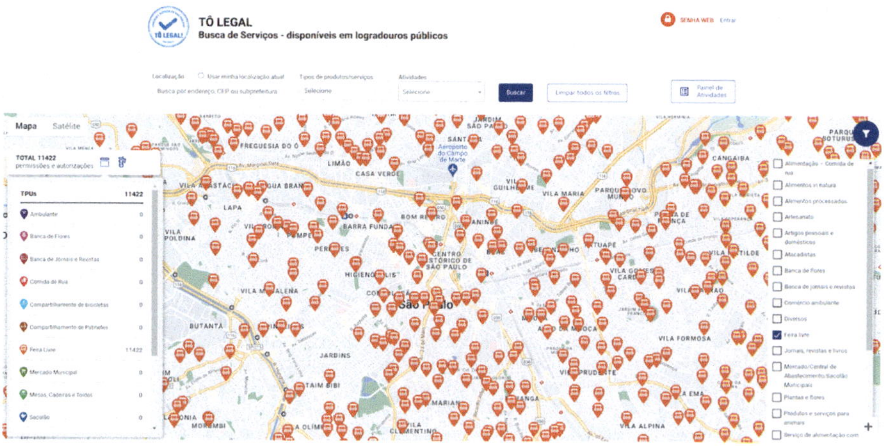

The above images are provided to us by the project owners for their submission.

Chapter 15
Substitution of Chemical Pesticides by Using Pruning Waste in the Production of Herbal Formulations in Tehran, Iran

Abstract The most significant impact of green spaces in cities lies in their environmental functions and ecological efficiency, which enhance urban livability. A study conducted in Bostan Sufi, within Tehran's Municipality Region 17, examined the effectiveness of several herbal essential oil formulations—derived from rosemary, peppermint, and eucalyptus—against the whitefly pest *Aleuroclava jasmini*. These natural solutions were compared to the chemical pesticide diazinon (Basodox 60%). Additionally, research was conducted on the use of the *Caesalpinia gilliesii* plant as a natural pest control method for urban pests such as the white-winged sparrow (*Siphoninus phillyreae*), black willow aphid (*Chaitophorus niger Mordvilko*), ornamental cabbage waxy aphid (*Brevicoryne brassicae*), rose bush aphid (*Macrosiphum rosae*), and marshmallow aphid (*Aphis gossypii Glover*). The scientific achievements of this research have been documented in five published articles. The outcomes of this project support the promotion of pruning plants within green spaces while encouraging domestic production through knowledge-based companies as an alternative to chemical pesticides. The plants utilized in this study include Arizona cypress, lavender, and rosemary, among others. This initiative aligns with Tehran's municipal strategy and the 2030 Sustainable Development Goals, emphasizing sustainable urban management and green space maintenance based on environmental sustainability principles. By adopting these organic solutions, Tehran is taking meaningful steps toward a cleaner and more ecologically responsible future. This initiative is related to SDGs 2, 3, 4, 7, 8, 9, 10, 11, 15, and 17.

Keywords Green spaces · Ecological efficiency · Urban pest control · Herbal essential oils · Sustainable development · Environmental sustainability · Organic pesticides

A. Cheshmehzangi et al., *People-Centered Innovation*, Urban Sustainability,
https://doi.org/10.1007/978-981-96-5230-3_15

15.1 Introduction to Tehran, Iran

Tehran, the capital and largest city of Iran (as well as the Middle East), stands as the country's political, economic, and cultural hub. With a population of approximately 9,013,700, it is the most populous urban center in Iran and one of the largest cities in the Middle East. Spanning an area of 730 km^2, Tehran has a high population density of 11,800 people per square kilometer, reflecting its status as a highly urbanized and rapidly growing metropolis. With an annual population growth rate of 1.20%, the city continues to expand both in size and economic influence, shaping Iran's future in various sectors.

Tehran is the economic engine of Iran's growth, making a substantial contribution to the GDP of the nation. With a GINI index of 0.4 and a GDP per capita of $4091, the city has a moderate degree of income inequality. Tehran is home to the country's largest corporations and governmental organizations, and its economy is broad, encompassing sectors like manufacturing, trade, technology, and finance. The slogan "From Nature, for Nature" reflects the city's primary source of prosperity, which is closely linked to Iran's natural resources and environmental sustainability initiatives. This strategy establishes Tehran as a pioneer in sustainable urban development and demonstrates its dedication to striking a balance between environmental responsibility and economic growth.

Tehran enjoys a distinct topographical and climatic advantage due to its location in the foothills of the Alborz Mountains. Because of its basin-like geography, the city has a continental climate with hot summers and cold winters, as well as significant air pollution issues. Tehran has been aggressively enacting laws pertaining to pollution prevention, green spaces, and sustainable infrastructure in order to address these environmental issues. Urban green spaces are essential to the city's environmental management because they improve the quality of life for locals while reducing air pollution.

Tehran is home to many museums, landmarks, and heritage sites that showcase Iran's rich past, making it a hub for history and culture. Theaters, art galleries, and music festivals that feature both traditional and modern Iranian art are all part of the city's thriving cultural scene. With some of the nation's best universities and research facilities, Tehran is also a significant hub for education, advancing the nation's knowledge and technology. Tehran has a number of urban problems, such as air pollution, traffic jams, and a lack of housing, despite its economic advantages. Nonetheless, the city is striving to create a more resilient and livable urban environment through environmental sustainability initiatives, public transportation investments, and creative urban planning. Tehran continues to develop as a vibrant city, influencing Iran's future while preserving its enduring ties to the natural world, thanks to a strategic vision that balances economic expansion with environmental stewardship.

15.2 Background and Origins

The HSE-WI-03-01 article highlights the significance of preserving urban green spaces while guaranteeing the welfare of citizens by outlining the fundamental concepts of health, safety, and environmental services in these areas. The reduction of health and environmental hazards linked to pesticide use is a fundamental component of this framework, which promotes more ecologically friendly methods of pest management. Sustainable green space management is becoming more and more important as urban areas expand, necessitating creative solutions that complement ecological conservation initiatives and public health priorities.

The mulberry whitefly outbreak in Tehran's urban green spaces between 2016 and 2018 presented a serious problem. Particularly impacted was Atrat Park, which is situated in the 17th municipality's second district. As a result, there is an immediate need for efficient pest management techniques. A crucial discovery resulted from this crisis: the *Caesalpinia gilliesii* plant naturally attracted adult whiteflies, making it a viable option for biological pest control. After extensive study, the plant's function as a "trap plant" was incorporated into the city's executive agenda for managing urban green spaces.

According to observations made by the department of herbal medicine, *Caesalpinia gilliesii's* glassine and fruit (seeds) contained toxic trichomes that effectively killed white-winged insects. The city's dedication to environmentally friendly pest control techniques was strengthened by this discovery, which offered a sustainable substitute for chemical pesticides. The results demonstrated the potential of plant-based approaches to managing urban ecosystems and opened the door for more research on how botanical insect control techniques can lessen dependency on artificial chemicals.

The wider ramifications of this study went beyond Atrat Park and affected Tehran's urban sustainability regulations. Municipal officials sought to improve the general health of urban ecosystems, preserve biodiversity, and lessen environmental contamination by incorporating such natural solutions into green space management. This change was also in line with worldwide patterns for sustainable urban growth, which aim to strike a balance between ecological stewardship and modernization. By proving that nature-based solutions could be both practical and effective in expansive urban settings, the *Caesalpinia gilliesii* project became a model for innovation in pest control.

The knowledge gained from this project is still essential for future urban planning as Tehran grows. The effectiveness of employing *Caesalpinia gilliesii* as a trap plant emphasizes the value of multidisciplinary research, departmental cooperation within the municipality, and funding for sustainable solutions. Additionally, it emphasizes the need for proactive environmental policies that put long-term ecological health ahead of immediate benefits. Tehran is making significant progress toward greener, healthier, and more resilient urban areas by adopting nature-inspired practices.

15.2.1 Key Innovative Aspects

This is the first time the *Caesalpinia gilliesii* plant has ever been used for this purpose globally, and its use in the creation of herbal formulations is a revolutionary innovation. This innovative strategy is the result of in-depth investigation and cooperation between the nation's herbal research institute and Tehran's 17th municipal district, underscoring a special alliance between local government and academic institutions. This collaborative effort facilitated the project's funding and implementation, allowing for a thorough investigation of the plant's characteristics and potential as a natural pesticide.

The creation of plant-based pesticides has been a top priority for scientists and research groups for a number of years in an effort to lessen environmental damage and encourage the production of healthier agricultural products. This program supports international initiatives to reduce the use of chemical pesticides and promote sustainable urban and agricultural practices. Following a series of scientific and technical exchanges, *Caesalpinia gilliesii*-derived primary formulations were sent to Tehran's 17th municipal district for carefully monitored in vitro and in vivo testing. These tests were crucial in confirming the plant's efficacy as a biological pest control technique and honing its use in managing urban green spaces.

The project included thorough statistical analyses in addition to laboratory studies to evaluate the effectiveness of herbal formulations versus traditional chemical treatments. The study's encouraging findings supported *Caesalpinia gilliesii's* potential as a sustainable substitute for artificial pesticides. Through the publication of research articles and reports by the Tehran Studies and Planning Center, these findings not only helped shape local environmental policies but also added to the larger scientific conversation. The project's legitimacy has been further strengthened by the registration of these scientific results, which has also established a standard for further study in the area of plant-based pest control.

This initiative's interdisciplinary approach, which blends state-of-the-art botanical research with urban environmental management, is among its most inventive features. A model for cooperative problem-solving has been established by the combination of scientific knowledge and municipal governance, showing how research-driven policies can result in workable, scalable solutions. Tehran has made a big move in addressing urban ecological issues and advancing sustainability by investing in plant-derived pest control techniques. In addition to improving the city's green areas, the project's success sets an example for other towns looking to adopt eco-friendly pest control techniques.

The significance of ongoing research and development in the field of herbal pest control is highlighted by this initiative. The knowledge acquired from using *Caesalpinia gilliesii* emphasizes the possibility of more research into plant-based remedies in both urban and rural contexts. Future policies in the city will be influenced by the lessons learned from this creative project, which will promote the adoption of more sustainable and scientifically supported methods of managing the urban environment. Tehran has established itself as a leader in environmentally sustainable urban development with this innovative endeavor, opening the door for other cities to adopt similar strategies by incorporating natural solutions into their urban planning frameworks.

15.2.2 Desired Outcome and Measurement

15.2.2.1 Desired Outcome

By growing and producing biocompatible pesticides from *Caesalpinia gilliesii* and other plant-based formulations in large quantities, the project aims to create an environmentally responsible and sustainable method of controlling urban pests. The project's goal is to develop safe, efficient substitutes for chemical pesticides by using natural resources, such as the leftover pruning of evergreen plants like lavender and rosemary. By reducing the negative effects of synthetic pesticides on the environment and public health, this initiative is anticipated to make a substantial contribution to urban health.

One of the main goals is to incorporate these plant-based formulations into Tehran's larger urban management plan, which will guarantee that green areas are kept up properly and lessen reliance on dangerous chemicals. The municipality aims to improve biodiversity, strengthen urban ecosystems' resilience, and create a cleaner, healthier city by implementing nature-based solutions. This change is anticipated to have major economic benefits in addition to environmental ones. By offering a cost-effective and sustainable pest control solution, the growth of regional, sustainable pesticide manufacturing will assist in lowering maintenance expenses for local government agencies, ultimately resulting in long-term financial savings.

The project's goal is to support scientific research and innovation in sustainable urban management, which is another crucial component. Tehran hopes to establish a standard for other cities wishing to adopt comparable environmentally friendly methods by expanding our understanding of plant-based pest control. Additionally, the project aims to incentivize local businesses and knowledge-based enterprises to invest in the manufacturing and marketing of herbal formulations. The initiative seeks to establish a self-sustaining system that supports ongoing advancements in urban environmental management by encouraging cooperation between private companies, research institutions, and municipal authorities.

The successful commercialization of plant-based formulations, decreases in the use of chemical pesticides, increases in the effectiveness of urban pest control, and savings for municipal maintenance are just a few of the important metrics that will be used to gauge success. The project will also monitor the environmental impact, evaluating elements like the health of green spaces and enhancements in biodiversity. Benefits to public health, such as lower risks of pesticide-related illnesses, will also be a crucial indicator of the initiative's long-term success. Tehran hopes to set an example by accomplishing these goals and proving that nature-based, sustainable solutions can be successfully incorporated into contemporary urban management.

15.2.2.2 Measurement

To ensure a thorough assessment of the project's impact, a combination of scientific, economic, and social indicators are used to measure its success. When evaluating the effectiveness of plant-based formulations made from *Caesalpinia gilliesii* and pruning waste from evergreen plants, biometric tests are essential in both basic and agricultural sciences. Advanced analytical tools like POLO-PC, SPSS, and Design Expert software are used in these assessments to produce accurate data on how well the formulations control urban pests while preserving ecological balance. By using these techniques, scientists can confirm that plant-derived pesticides are effective, stable, and long-lasting, making sure they fulfill all safety and performance requirements prior to widespread use.

Apart from scientific evaluations, economic performance is a crucial metric. The project's ability to establish a self-sufficient biocompatible pesticide production industry is assessed by looking at how profitable it is to use pruning waste as a raw material. In order to lessen dependency on pricey synthetic pesticides, the objective is to guarantee that knowledge-based businesses and local governments can incorporate these organic formulations into their pest control plans at a competitive price. The project's economic viability and long-term viability are assessed using financial metrics, such as the potential revenue from commercialized plant-based formulations, the decrease in the cost of purchasing chemical pesticides, and the savings in municipal maintenance.

Another crucial aspect of the evaluation process is social impact. By reducing exposure to dangerous chemical pesticides, the project seeks to protect the health of both residents and municipal employees. Health-related data, such as the prevalence of respiratory disorders and pesticide-related illnesses among city workers and residents, are tracked over time to gauge its efficacy. To make sure that residents and local managers understand the advantages of switching to more environmentally friendly pest control techniques, public opinion polls and community involvement programs aid in determining the level of awareness and acceptance of plant-based substitutes.

Additionally, improvements in biodiversity within Tehran's urban green spaces are monitored by environmental impact assessments. Researchers can assess how well plant-based formulations support a healthy and balanced urban ecosystem by examining soil quality, vegetation health, and pest population control effectiveness. Other ecological indicators of success include improvements in air quality, a decrease in chemical pesticide runoff, and an increase in populations of beneficial insects like pollinators. The incorporation of these measurement tools strengthens the project's role in advancing sustainable urban environmental management by ensuring that it stays data-driven, transparent, and flexible to future developments.

15.2.3 *Implementation*

15.2.3.1 Principal Parties and Partners

Collaboration between a number of important parties, including public institutions, private sector stakeholders, research organizations, and urban society, has been essential to the project's successful execution. As part of Tehran's larger urban pest control plan, the municipal organization has been instrumental in the widespread planting of *Caesalpinia gilliesii*, a natural whitefly trap. The municipality has made an effort to lessen its dependency on chemical pesticides by giving preference to eco-friendly alternatives. This has improved public health and reduced maintenance expenses for the municipality.

The Institute of Plant Medicine Research has been a vital collaborator in this endeavor, helping to develop and validate plant-based pesticides scientifically. Because of this partnership, thorough research and testing have been made possible, guaranteeing that the herbal formulations are safe and effective enough for widespread use in urban green spaces. A sustainable substitute for traditional pesticides has been made possible by the collaboration between the research institute and the municipality, which has enabled the creation of creative solutions suited to the city's economic and environmental requirements.

The goals of the project have also been greatly aided by the private sector. The development and possible commercialization of these plant-based solutions have been carried out by knowledge-based businesses that specialize in herbal formulations and biocompatible pesticides. These businesses support a circular economy model by turning organic waste into useful pest control products by using pruning waste from evergreen plants like lavender and rosemary. Private companies' participation guarantees the project's continued economic viability and opens the door for its potential to spread beyond municipal applications into larger agricultural and urban management markets. Another important stakeholder has been urban society, which includes community organizations and environmental organizations. Their backing has been essential in promoting environmentally friendly urban pest management techniques and increasing public knowledge of the advantages of cutting back on the use of chemical pesticides. The project has ensured long-term public support for nature-based solutions by fostering a sense of shared responsibility for environmental health through citizen engagement.

These varied partners' combined efforts have enabled Tehran's municipal organization to put into practice a creative pest control plan that is supported by research and the community. The city's dedication to environmental sustainability and public health has been strengthened by this multi-sectoral partnership, which has also resulted in significant cost savings. The project serves as a model for combining community involvement, private sector innovation, municipal governance, and scientific research in urban environmental management with the ongoing collaboration of these key players.

15.2.3.2 Resources for Implementation

This project's successful execution has required a mix of technical, financial, and human resources. The project will be guided by specialized knowledge and scientific rigor thanks to the careful selection of human resources from among herbal medicine experts. These experts, who are mostly from the national Herbal Medicine Institute, have been in charge of assessing how well plant-based formulations work to control urban pests. Their proficiency has been essential in carrying out field tests, laboratory experiments, and data analysis to confirm the formulations' efficacy prior to their extensive implementation in Tehran's green areas.

The Municipal Organization has been the main source of funding for the project because it has seen the long-term advantages of switching from chemical pesticides to more economical and sustainable alternatives. The initiative has been funded in a number of ways, including laboratory research, pilot testing, plant material procurement, and project staff training. The publication of research findings has also been aided by financial resources, guaranteeing that the findings are recorded, shared, and available for additional scientific investigation and real-world use. The project has built a solid foundation for long-term sustainability and possible growth by obtaining municipal funding.

The project's implementation has also been greatly aided by technical resources. To perform biometric tests and assess the effectiveness of plant-based pesticides, sophisticated laboratory equipment has been employed. By using statistical analysis and modeling software like POLO-PC, SPSS, and Design Expert, researchers have been able to evaluate the effects of herbal formulations in a variety of scenarios. These resources have given decision-making a scientific foundation, guaranteeing that the formulations are both practical and profitable. In addition to research and testing, implementation has necessitated incorporating the developed pesticides into the management of urban green spaces. In collaboration with municipal landscaping teams, *Caesalpinia gilliesii* has been planted widely as a natural whitefly trap. The plants are positioned to optimize their pest control advantages. To further improve sustainability, the project has also made use of already-existing municipal infrastructure to collect and use pruning waste from evergreen plants like lavender and rosemary.

This initiative has successfully moved from research to implementation thanks to a combination of highly qualified human resources, committed municipal funding, state-of-the-art research instruments, and strategic urban planning. The project has shown a model for incorporating ecological pest control techniques into urban management by coordinating financial, scientific, and operational resources, opening the door for more widespread applications in sustainable urban development.

15.2.3.3 Tools and Methods Applied

This project's execution has been directed by a set of precise instruments and techniques meant to incorporate plant-based formulations into urban pest management plans. The strategy is based on scientific research, environmental sustainability, and cooperation between knowledge-based businesses, research institutes, and local government agencies. The controlled use of *Caesalpinia gilliesii* in pilot gardens to control whiteflies has been one of the main strategies used. The plant's ability to draw in and repel whiteflies has been methodically observed and documented in these pilot areas, which have functioned as experimental locations. This strategy has ensured that the techniques created are sustainable and effective by offering vital information on the viability of large-scale implementation.

Using specially created plant-based formulations to manage aphid pests in urban green spaces, like the Salamat Park family farm, has been another significant strategy. Scientists have been able to fine-tune the composition of pesticides derived from plants by developing these formulations through a combination of laboratory research and field testing. The objective has been to develop biocompatible pest control methods that are highly effective at preserving urban plant life while lowering the need for artificial chemicals. Testing various plant extracts, including those made from lavender and rosemary, has been part of the process to see how well they work against different urban pests.

To guarantee the appropriate management of pruning waste, the project has created a structured environmental program in accordance with municipal regulations, in addition to field experiments. A specific circular on urban health states that in order to make it easier to collect and use pruning waste from seasonal trimmings, agreements have been established between research institutes, knowledge-based businesses, and municipal authorities. In order to create a closed-loop system where organic materials are used for pest control rather than being thrown away, this waste is then processed into plant-based formulations. In addition to improving sustainability, this approach helps urban landscaping reduce waste and use resources more efficiently.

Relying on chemical sciences to improve and refine plant-based formulations has been a crucial scientific approach used in this project. To guarantee the efficacy and safety of the created pesticides, researchers have used controlled application trials, chemical analysis, and biometric testing. The industry that produces pesticides has adopted these formulations, and the acceptance and cultural assimilation of these formulations have been greatly aided by municipal authorities. Incorporating biocompatible pest control techniques into urban management practices has been made easier by the project's promotion of cooperation between researchers and policymakers.

The implementation of the project has been further aided by the use of digital tools. Field data analysis, pest behavior modeling, and formulation efficiency optimization have all been accomplished with the use of sophisticated statistical software like POLO-PC, SPSS, and Design Expert. These resources have made it possible for researchers to substantiate their conclusions with solid scientific data, guaranteeing that the formulations created satisfy the strictest performance and safety requirements. The project has effectively demonstrated an inventive, sustainable, and scientifically based approach to urban pest control using these well thought-out tools and techniques. Setting the standard for upcoming ecological projects in Tehran and elsewhere, the incorporation of plant-based formulations, waste management techniques, and digital analysis tools has created a more ecologically friendly urban environment.

15.3 Obstacles and Solutions

There have been a number of obstacles to the successful application of plant-based pesticides in Tehran's urban green areas, which call for calculated fixes. The cultural reluctance to use biocompatible pesticides has been one of the main challenges. Chemical pesticides have historically dominated pest management, so a mental shift among urban planners, local government representatives, and the general public is necessary to make the switch to plant-based alternatives. Because of their widespread use and perceived efficacy, many stakeholders have been reluctant to stop using chemical treatments.

In order to address this, widespread awareness campaigns have been carried out to inform citizens, environmental authorities, and municipal employees about the advantages of biocompatible formulations. By emphasizing the positive outcomes of pilot programs, these initiatives have highlighted the benefits to the environment and human health of lowering the use of chemical pesticides. The unwillingness of certain organizational managers to believe in the efficacy of herbal formulations, even in the face of evidence showing their success, has been another significant obstacle. The ability of plant-based remedies to control urban pests as effectively as synthetic pesticides has been questioned by some officials. This opposition has restricted the project's funding allocation and delayed its broad adoption. Thorough scientific validation, including controlled field experiments and biometric testing, has been given top priority in order to address this. Research articles published in reputable scientific journals have gained additional legitimacy, confirming the efficacy of plant-based pesticides and persuading more decision-makers to take them into account as a feasible substitute.

Another major obstacle has been financial constraints. Although the municipal government has expressed interest in eco-friendly projects, it has proven difficult to obtain steady funding for knowledge-based businesses engaged in the domestic manufacturing of plant formulations. Bureaucratic procedures and some municipal bodies' reluctance to make long-term investments in plant-based pest control solutions have caused delays in numerous funding proposals. Partnerships with environmental groups and private sector investors have been investigated as a way to deal with this, guaranteeing a variety of funding sources. The project has also been prioritized in municipal planning and budget allocation, with attempts made to incorporate it into more comprehensive urban sustainability policies.

Another obstacle has been the absence of official agreements between the municipality and knowledge-based businesses regarding the use of pruning plants in the manufacturing of household pesticides. The possibility of increasing production and ensuring a consistent supply of plant-based pesticides is unclear in the absence of formal contracts. In order to ensure a solid foundation for collaboration between research institutions, environmental organizations, and urban management bodies, advocacy efforts have been focused on municipal authorities to institutionalize these partnerships. Formalizing these partnerships and guaranteeing the project's long-term viability require clear policy directives and regulatory support.

Despite these challenges, the project has effectively addressed each one by implementing strategic solutions. Adoption of plant-based pesticides in Tehran's green spaces has been aided by policy advocacy, financial diversification, scientific validation, and awareness campaigns. By showing that environmental innovation can be incorporated into municipal governance with the correct mix of research, education, and policy support, these initiatives have set the stage for a long-term shift toward sustainable urban pest management.

15.3.1 Achieved Outcome

Tehran's urban green spaces have benefited greatly from the use of plant-based pest control techniques, especially in terms of lowering dependency on chemical pesticides. The successful integration of the *Caesalpinia gilliesii* plant as a natural trap for whiteflies, one of the most enduring urban pests, has been a major accomplishment of this initiative. This plant, which provides an environmentally friendly substitute for chemical treatments, has been shown through extensive research and field trials to be very effective in attracting and controlling white-winged flying insects. This plant's ability to reduce pest-related issues has been shown by its widespread cultivation in city parks and green areas, especially during the hottest months when whiteflies proliferate quickly and become resistant to artificial pesticides.

The *Caesalpinia gilliesii* plant was extensively tested as a trap crop in a number of pilot gardens between 2016 and 2017. By acting as a natural attractor, the plant's vivid yellow blooms greatly decreased the number of whiteflies in the nearby vegetation. Due to its success, this technique was extended to several urban parks, where it was demonstrated that it could effectively manage whitefly populations without the use of chemicals. As a result, the city saw a significant drop in the use of pesticides, which helped to protect the environment and reduced the health risks that residents and municipal employees faced from extended exposure to harmful substances.

The project's improvement of urban biodiversity and ecosystem balance has been another noteworthy result. The trap plant method enables more focused pest control while maintaining ecological integrity, in contrast to chemical pesticides, which frequently upset natural predator–prey relationships by eradicating both pests and beneficial insects. Beneficial insect populations, like pollinators and natural predators, have increased as a result of this strategy. These insects are essential to the upkeep of healthy green spaces.

The municipal government has also benefited financially from the decrease in the use of chemical pesticides. Maintenance expenses for urban parks and green areas have decreased due to a decreased reliance on costly chemical treatments. Tehran's dedication to reaching long-term environmental goals has been strengthened by the use of biocompatible pest control solutions, which have opened the door for additional investments in sustainable urban management techniques. Further investigation into using pruning waste from evergreen plants, such as lavender and rosemary, for the creation of other plant-based formulations has also been spurred by the initiative's success.

The project has contributed to Tehran's larger environmental sustainability agenda in addition to providing immediate pest control. The municipal government's investment in environmentally friendly solutions supports Tehran's attempts to fulfill its international environmental obligations and is in line with global trends in urban sustainability. The project serves as a model for other cities looking to lessen their chemical footprint while preserving urban greenery by showcasing the large-scale efficacy of plant-based pest control. The accomplishments of this project demonstrate the potential of using natural solutions to manage cities. Tehran has established a standard for future ecological interventions in urban settings by successfully integrating a sustainable substitute for chemical pesticides through scientific innovation, strategic implementation, and policy support. The city's experience with this project emphasizes how crucial it is to integrate research and real-world implementation to make sure that initiatives for urban sustainability are both successfully carried out and supported by science.

15.4 Lessons Learned

Insights into sustainable pest management have been gained from the use of biocompatible pesticides in Tehran's urban green areas. The significance of scientific validation and documentation is one of the project's main lessons. Promoting the use of herbal pesticides, guaranteeing their legitimacy, and fostering wider acceptance within the agricultural and urban management sectors have all benefited greatly from the registration of research articles and scientific findings. The project has helped to increase the body of scientific knowledge regarding sustainable pesticide alternatives and urban pest control by methodically recording results.

The need for institutional cooperation and regulatory support is another important lesson. This initiative's success has shown that cooperation between research institutions, agricultural agencies, and municipal organizations is necessary to guarantee the successful adoption and expansion of such innovations. Although agricultural organizations' eagerness to disseminate herbal formulations has been a good thing, there were initially difficulties due to a lack of general awareness and regulatory alignment. An important step toward institutionalizing these practices is the development of new regulations targeted at lowering the use of chemical pesticides in greenhouses, gardens, and municipalities. It is possible to guarantee the long-term viability of the use of biocompatible pesticides by cultivating greater legislative support.

Another important lesson learned from this project is economic viability. It has been shown that lowering dependency on chemical pesticide imports is both economically and environmentally advantageous, particularly for nations subject to trade restrictions or embargoes. Encouraging the domestic production of herbal pesticides can save a lot of money while maintaining the quality of urban green spaces and agricultural products. In order to create alternatives that are both economical and ecologically friendly, this experience emphasizes the necessity of ongoing investments in regional research and production capacities.

The significance of cultural acceptance and public awareness has also been emphasized by this project. Because of their long-standing reliance on chemical pesticides, some organizational managers initially opposed the adoption of herbal formulations, despite their demonstrated effectiveness. This illustrates that in addition to scientific innovation, successful implementation necessitates educational initiatives to change attitudes and highlight the advantages of sustainable pest control techniques. Achieving long-term adoption requires involving important stakeholders and decision-makers in the change process. Ultimately, this project has confirmed the potential of using natural solutions to manage cities. Tehran's experience with biocompatible pesticides serves as a model for other cities looking to shift toward more sustainable urban practices by combining scientific research, policy support, economic considerations, and cultural awareness. The project's lessons emphasize the need for a comprehensive strategy that strikes a balance between social, economic, and environmental aspects in order to successfully adopt green innovations.

Related SDGs: Goals 2, 3, 4, 7, 8, 9, 10, 11, 15, and 17.

Acknowledgements We explicitly thank the project owners based in Tehran, Iran, for their participation and the provision of materials for the 6th Guangzhou International Award for Urban Innovation. We also congratulate them on their achievements in the 2023 cycle of awards.

The above images are provided to us by the project owners for their submission.

Chapter 16
Enhancing Traffic Safety Near Schools Through Child-Friendly Collaborative Construction and Sharing' Approach in Xianning, China

Abstract In China, traffic congestion and safety concerns near schools pose significant challenges for urban management, particularly in medium-sized cities like Xianning. Many parents, apprehensive about their children's safety, opt to drive them to school, exacerbating traffic issues, increasing air pollution, and leading to considerable time wastage. However, financial constraints and the absence of national directives make it difficult for local governments to implement effective interventions. Addressing this issue requires a holistic, child-friendly, and collaborative approach that integrates urban planning, community participation, and institutional cooperation across multiple departments, including education, transportation, and law enforcement. Additionally, fostering behavioral change through awareness programs for parents, students, and local residents is critical to ensuring long-term success. In order to promote shared mobility solutions, safe zones, and better road infrastructure surrounding schools, this study investigates a novel "Collaborative Construction and Sharing" model that is specifically designed for Xianning. This strategy seeks to make school transportation safer, more sustainable, and more accessible by utilizing neighborhood resources and involving stakeholders. The results offer scalable and flexible solutions for child-friendly urban mobility, which is helpful for other medium-sized cities dealing with comparable issues. This initiative is related to SDGs 3, 4, 5, 11, 13, 15, and 16.

Keywords Traffic safety · Child-friendly urban planning · Collaborative construction · Shared mobility · School commuting · Medium-sized cities · Stakeholder engagement

16.1 Introduction to Xianning, China

In China's southeast Hubei Province, the prefecture-level city of Xianning is renowned for its abundant natural resources, expanding economy, and advantageous location along the Yangtze River Economic Belt. With 3,036,100 residents and a consistent 0.77% annual growth rate, Xianning has managed to strike a balance

A. Cheshmehzangi et al., *People-Centered Innovation*, Urban Sustainability,
https://doi.org/10.1007/978-981-96-5230-3_16

between its agricultural heritage and urban growth. With a population density of 268 persons per square kilometer and a total area of 9752 km^2, the city exhibits moderate urban development in comparison to other significant Chinese cities.

With a GDP per capita of $8590, Xianning's economy is mainly fueled by tourism, agriculture, and public health services. The city is a major producer of tea, rice, and products related to bamboo because of its fertile land and ideal climate, which support its thriving agricultural sector. Xianning's status as the "City of Hot Springs," which draws tourists to its many thermal spa resorts, picturesque mountain ranges, and cultural heritage sites, boosts the travel and tourism sector. Furthermore, public health services—with a particular focus on traditional Chinese medicine and wellness-based industries—are essential to the growth of the city. Xianning's GINI index of 0.46 indicates that, despite its economic growth, the city still has issues with income inequality. The city is making investments in social programs, education, and infrastructure to increase opportunities for lower-income and rural residents in an effort to address disparities.

Sustainability has been incorporated into Xianning's urban development plans in recent years. The city has aggressively supported green initiatives, such as water conservation projects, reforestation programs, and the growth of ecotourism. The vast forested areas of Xianning are being preserved in accordance with national policies on biodiversity conservation and carbon reduction. Another major area of focus is the development of transportation infrastructure, with ongoing initiatives to enhance road safety, pedestrian-friendly areas, and public transportation. Urban planning in Xianning places a strong emphasis on striking a balance between environmental preservation and modernization, especially when it comes to controlling traffic around commercial areas and schools. Projects to improve road networks, support shared mobility options, and promote non-motorized transportation options have been started by the city's authorities. Xianning is dedicated to sustainability, which includes pollution prevention and waste management. In addition to investing in renewable energy sources like solar power and biomass energy made from agricultural waste, the city has tightened environmental regulations for industrial operations. These programs support improved air and water quality while lowering the city's carbon footprint.

In order to attain balanced growth in the future, Xianning hopes to capitalize on its advantages in ecotourism, sustainable agriculture, and healthcare services. New urban planning models that put an emphasis on community involvement, smart city technologies, and climate resilience are being investigated by policymakers. One of the most important aspects of these initiatives is addressing traffic safety, especially in the vicinity of schools, which calls for creative solutions and cooperative governance. Xianning serves as a model for medium-sized Chinese cities looking to combine economic growth with environmental and social well-being because of its unique combination of historical significance, scenic beauty, and progressive policies. The city's future course will be greatly influenced by its approach to inclusive urban planning and sustainability as it develops further.

16.2 Background and Origins

Traffic congestion around schools in Xianning has become a pressing urban challenge, particularly due to the limited space allocated for school drop-offs and pick-ups. The daily "school run" leads to chaotic traffic conditions, compromising both safety and efficiency. Economic disparities across different districts further exacerbate the issue, as some schools are better resourced than others, leaving students in disadvantaged areas exposed to unsafe commuting conditions.

The 7th Elementary School of Xianning, which is situated in a suburban area with inadequate road infrastructure, is a perfect illustration of this problem. Basic facilities like streetlights, security cameras, and secure pedestrian walkways are still absent from the school's access roads, which are still in bad shape. Teachers and staff are forced to manually control traffic at a crucial intersection close to the school because there are no traffic signals, which puts both students and teachers in danger. Despite these persistent issues, the lack of a coordinated, citywide intervention and a lack of funding have caused the necessary improvements to be postponed.

The project to improve traffic safety around schools through child-friendly collaborative construction was started because it was realized that immediate action was required. The main goal of this project is to turn undeveloped school grounds into attractive, welcoming, and educationally stimulating public areas. The program seeks to make school commuting safer and more interesting by enhancing pedestrian infrastructure, establishing safe zones, and incorporating urban design that puts children's needs first. Additionally, by creating environments that promote play, creativity, and active engagement in urban life, the initiative highlights the significance of social interactions and cognitive development among children, especially those from marginalized backgrounds.

The national and international child-friendly urban development policies are in line with this endeavor. The initiative is based on the UN's 1996 "Child-Friendly Cities" concept and adheres to recent Chinese directives such as the "Guiding Opinions on Promoting the Construction of Child-Friendly Cities" and the "China Children's Development Program (2021-2030)." All children, regardless of socioeconomic background, will have access to safe, green, and well-structured public spaces thanks to the initiative's emphasis on equitable urban opportunities, which incorporates principles from the New Urban Agenda and the Sustainable Development Goals (SDGs). By proving that child-friendly infrastructure can act as a catalyst for the development of healthier, more inclusive, and more sustainable urban environments, the larger objective is to transform Xianning's approach to urban planning.

16.2.1 Key Innovative Aspects

The creative and child-centered approach of the Xianning traffic safety improvement project near schools sets it apart. This project's dedication to designing from a "one-meter height" perspective—literally, viewing the world from a child's point of view—is among its most distinctive features. In close collaboration with students, engineers, urban planners, and designers actively consider their suggestions and integrate their originality into the planning process. As a result, children's creative input directly informs design features like wall murals and themed areas, like a "Protecting the Earth" area. In addition to making the school environment safer, this strategy gives students a sense of empowerment and ownership.

One of the initiative's main pillars is public participation. The project guarantees that urban spaces are not only useful but also entertaining and instructive by incorporating kids, parents, educators, and locals in the renovation of school environments. In order to engage the community and students in urban greening initiatives, abandoned areas near schools have been transformed into shared gardens. These areas have two functions: they offer outdoor learning opportunities in addition to green, tranquil settings. This cooperative model promotes an inclusive approach to urban development and improves social cohesion.

The implementation of a franchise policy to eliminate physical barriers in urban areas is one of the initiative's most innovative tactics. In order to expand common areas and incorporate them into a shared urban fabric, real estate developers have been encouraged to tear down the boundary walls that enclose residential complexes. Because of this policy, there are now more open and accessible spaces where businesses that deal with dining, recreation, and after-school activities can flourish. This promotes neighborhood interaction and child-friendliness while also bolstering local economic development.

A fundamental component of this program is sustainability, which emphasizes the encouragement of eco-friendly, low-carbon, and green lifestyles. For example, the "Shared Garden" project incorporates sustainability principles into the educational process by using building materials made from used bottles and cans that children have gathered. Over half of the materials used in the construction of public areas surrounding school campuses have been recycled, which lessens the impact on the environment while also teaching children and their families/parents/guardians about waste reduction and climate-conscious behavior. Students are guaranteed to develop an innate awareness of ecological responsibility as a result of the initiative's incorporation of sustainability into daily life.

The development of a multi-sectoral governance framework, which guarantees seamless coordination among diverse stakeholders, is another important innovation of this endeavor. The project has created a collaborative governance model that enables more effective policy implementation and infrastructure upgrades by strengthening ties between seven important government sectors, such as public security, housing and construction, urban management, and education. Incorporating school administrators and community representatives into this network enhances decision-making procedures and guarantees that interventions are customized to meet local needs.

Four pedestrian-only school routes have been implemented as part of the initiative to improve pedestrian safety, which has greatly reduced traffic during peak hours. More than 100 parent volunteers have helped with this endeavor, helping to control traffic and making sure children can get to and from school safely. In addition to making urban areas safer and more pedestrian-friendly, this model has been shown to be very successful in strengthening a sense of community involvement.

This initiative's quick scalability is what makes it unique. The success and adaptability of the project are demonstrated by the fact that the practices created in Xianning have already been implemented in 20 more schools. Additionally, Xianning's future urban development will be in line with inclusive, sustainable, and child-friendly planning principles thanks to the establishment of citywide design guidelines and implementation standards. The initiative establishes Xianning as a model for other medium-sized Chinese cities looking to incorporate safety, sustainability, and participatory design into their urban planning frameworks by institutionalizing these best practices, thereby laying the groundwork for long-term urban transformation.

16.2.2 Desired Outcome and Measurement

16.2.2.1 Desired Outcome

This initiative's main objective is to greatly increase traffic safety around schools in Xianning, thereby improving the commute experience for students. The project aims to make the area safer and more enticing for students to walk to school by implementing improved road infrastructure and pedestrian-friendly urban design. The project also seeks to reduce traffic congestion, vehicle emissions, and the need for parents to drive their children every day, all of which will help create a more sustainable urban environment. It is anticipated that mobility behavior will fundamentally change, creating a culture in which students prefer walking and bicycling as forms of transportation.

The initiative aims to promote systemic change at the municipal level in addition to the immediate effects on traffic safety. School traffic safety is acknowledged as a top priority in urban planning and policy creation when the issue is brought to the attention of city-wide governance structures. Important government agencies, such as the Housing and Urban-Rural Development Bureau, the Urban Planning Department, the City Management Committee, the Traffic Police, Neighborhood Committees, and the Education Bureau, have already endorsed the initiative. In order to guarantee that school environments are created with children's safety and wellbeing in mind, the initiative aims to institutionalize this multi-sectoral approach and make child-friendly urban development standard practice throughout city planning processes.

The creation and distribution of standardized guidelines that can be duplicated in other cities and schools is one of the initiative's main results. These recommendations offer a methodical framework for putting child-friendly urban design into practice, along with best practices for enhancing public spaces, pedestrian zoning, road infrastructure upgrades, and traffic management techniques. In order to guarantee that traffic safety interventions are not restricted to a small number of pilot schools but rather become an essential component of larger urban planning initiatives, the initiative seeks to develop these comprehensive standards in order to establish a scalable model that can be implemented throughout Xianning and beyond.

Promoting increased community involvement in urban development is another important goal. The program aims to foster a sense of collective responsibility for school safety by including parents, educators, students, local companies, and neighborhood committees in decision-making processes. In addition to increasing the efficacy of the interventions, this participatory approach fortifies social bonds and promotes sustained civic engagement in preserving secure and sustainable school environments. The project also aims to change the way school zones are used and viewed over the long run. School surroundings can transform from simple transit hubs into lively community centers where children can participate in social activities, physical exercise, and creative learning by incorporating safe routes, shared public spaces, and greener surroundings. Children's independence, problem-solving abilities, and environmental awareness are all enhanced by this reimagining of school neighborhoods.

The ultimate goal of this project is to establish Xianning as a model of child-friendly urban design in China. The project intends to encourage comparable changes in other medium-sized cities by showcasing the value of inclusive participation, sustainable design, and collaborative governance. This will guarantee that children's safety and wellbeing continue to be at the forefront of urban development initiatives across the country.

16.2.2.2 Measurement

Through a combination of quantitative and qualitative assessment methods, the initiative's success is being methodically measured, guaranteeing a thorough analysis of its effects on community satisfaction, traffic safety, and urban functionality. The evaluation process, which is headed by the design team and the project implementation department, is set up to record both short-term and long-term changes brought about by the initiative. The shift in the degree of satisfaction among locals, parents, and students is a crucial indicator in this evaluation. Surveys and interviews were carried out prior to the project's implementation in order to determine baseline opinions regarding accessibility, traffic safety, and the general school commute experience. Similar evaluations are being conducted to measure changes in public opinion after the improvements. These surveys offer important information about whether families feel safer and whether students feel more comfortable riding their bikes or walking to school. To learn more about how the change has affected social and economic interactions in the area, local companies and community members are also being consulted.

In order to ascertain whether traffic congestion has decreased in the vicinity of school zones, the project team is closely observing changes in traffic operation indicators in addition to subjective satisfaction levels. This involves tracking variations in traffic patterns, pedestrian flow, and travel durations during peak hours. The team can determine whether the initiative has successfully lessened the chaotic conditions of the daily "school run" and whether safer pedestrian pathways have improved students' mobility by examining data from traffic cameras, sensors, and manual observations. To make sure the intervention is directly resulting in a safer environment, special attention is paid to comparing accident rates and near-miss incidents before and after the project's implementation.

The data collection process is being broken down by gender, age, and income brackets to guarantee an equitable evaluation of the project's impact. The design team is able to comprehend how various demographic groups are impacted by the changes thanks to this thorough approach. For instance, it allows them to evaluate whether younger students feel as safe as older ones, whether parents and female students view the improvements differently than their male counterparts, and whether the project's results are distributed fairly among low-income families, who might have fewer transportation options. This degree of specificity guarantees that no population segment will be unintentionally left out or disadvantaged by the initiative.

To supplement numerical data, a visual documentation approach is also being used. The physical transformation of school zones is being documented through before-and-after photos, which offer a concrete depiction of the enhancements in public space design, pedestrian accessibility, and infrastructure. In addition to providing proof of advancement, these photos are a means of influencing decision-makers, interested parties, and the general public to talk about the significance of child-friendly urban design. The initiative makes sure that its impact is not only accurately measured but also continuously improved based on feedback from the real world by incorporating these multifaceted assessment techniques. In order to ensure long-term sustainability and broader urban transformation, the model will be further optimized and replicated in other cities and schools thanks to the insights gathered from this evaluation process.

16.2.3 Implementation

16.2.3.1 Principal Parties and Partners

This initiative's execution is a prime example of the effectiveness of cooperation between the public sector, private businesses, and the neighborhood. This group effort is inspired by the initial spirit of collaboration that resulted in the establishment of the 7th Elementary School of Xianning in 1998, when 73 private citizens and 19 local businesses banded together to provide funding for the school's construction. This fundamental sense of shared responsibility has been rekindled by the 2023 initiative, which has turned it into a model for engagement between the public, private sector, and community.

The municipal government is leading this effort, and the mayor's direct participation shows how committed the city is to the project. The local government put in a lot of effort to bring together various departments, including urban planning, education, public security, transportation, and city management, after realizing that enhancing school traffic safety necessitates a multi-sectoral approach. This high-level coordination made sure that logistical and regulatory issues were resolved effectively, opening the door for a quick and successful implementation.

By contributing land for the construction of safer bike lanes and pedestrian walkways, the private sector helped to overcome the spatial limitations surrounding schools. Beyond monetary donations, private developers took an active role in changing the urban environment to promote mobility that is child-friendly. The transformation of a whole street in a nearby residential estate into a pedestrian-only area, which removed traffic in the immediate area of the school, was one of the most revolutionary initiatives. In order to create an open and inviting space that improved visibility and safety for young children making their way to school, developers also took down fences and walls. The private sector showed an inventive and socially conscious approach to urban development by putting pedestrians' experiences first, especially those who are one meter tall.

The active participation of the local community was equally important to the project's success. In an effort to address the issue of the "school run," parents took the initiative to create four "walking bus routes." The percentage of students walking to school increased dramatically as a result of these organized walking groups, which were run by parent volunteers. Prior to the initiative, only 30% of students walked, but after it was put into place, that number rose to an astounding 95%. This change reinforced the importance of active mobility from a young age by reducing traffic congestion and promoting healthier habits among students.

The project successfully incorporated environmental awareness into the curriculum while also enhancing transportation. Along the newly created pedestrian routes, each class—teachers and students—took responsibility for maintaining the plants. Through the integration of practical sustainability practices into everyday school life, this initiative expanded learning beyond the classroom. In addition to enhancing the school's surrounding urban landscape, the incorporation of plant care into the biology curriculum promoted a deeper understanding of environmental stewardship. This initiative has developed a comprehensive model for improving school traffic safety by bringing together public and private organizations as well as the local community. In addition to addressing current issues, it promotes a long-term cultural change in favor of cooperative urban development. Xianning has shown how an inclusive, multi-stakeholder approach can result in significant, long-lasting improvements in urban environments, especially for the wellbeing of children, through shared responsibility and active engagement.

16.2.3.2 Resources for Implementation

The initiative's successful execution required a creative and varied combination of material, human, and financial resources that were raised by the local government, private developers, community members, and local businesses working together. The Xianning Housing and Urban-Rural Development Bureau and the Xianning Urban Management Committee supplied the vital funding required to carry out the project. The 7th Elementary School's immediate surroundings have been made safer, easier to access, and more inviting for students and their families thanks to the significant infrastructure improvements made possible by their investment. Important features like road improvements, traffic control system installation, and the construction of safer bike lanes and pedestrian walkways were all covered by this funding.

The removal of sidewalk parking, which had long been a significant source of traffic around the school, was one of the most significant interventions made possible by this funding. The city was able to make more space for pedestrians and lessen the disorderly atmosphere that frequently developed during busy school hours by removing these areas. In an effort to alleviate traffic, the city also launched a school bus service, which gives families a convenient and safe way to drop their children off at school. This reduces the number of cars in the area and encourages safer walking routes for kids.

The private sector demonstrated the effectiveness of corporate social responsibility in urban development by playing an equally significant role in guaranteeing the project's success. By voluntarily tearing down fences and barriers that previously blocked pedestrian access around the school area, private developers kindly supported the project. This collaborative effort helped open up the area and create a more welcoming atmosphere for students, in addition to making room for safer walkways. In order to reduce additional traffic-related stress in the area, the developers also donated empty plots of land that were temporarily transformed into school bus parking lots.

The success of the project was also greatly aided by local industries, especially those that supplied the raw materials needed to create public areas. Waste materials, including tires, metal, and bottles that had been discarded, were donated by several nearby factories and turned into inventive and useful street furniture and decorations for public areas. In addition to being recycled, which lessened their negative effects on the environment, these materials were also turned into creative, child-friendly designs that raised the newly created areas' visual appeal. Local artists and community members collaborated with the students to create these artistic installations, giving the children hands-on experiences and encouraging a sense of pride in the upgrades. This integration of art and sustainability further enriched the learning environment, reinforcing the educational goals of the initiative while promoting environmental consciousness.

Additionally, factories donated recycled soil made from water plant sludge, which was used in the landscaping of public areas, demonstrating the innovative use of local environmental resources. This environmentally friendly strategy made it possible to create green spaces that enhanced the school's surroundings while also aiding in stormwater management and supporting the city's efforts to maintain a sustainable environment. In conclusion, a combination of public funding, private sector support, and community involvement provided a wide range of abundant resources for the implementation of this initiative. One effective resource mobilization model that combined material, human, and financial contributions from multiple sectors was the city of Xianning. This cooperative strategy made sure that the project not only addressed the pressing issues of traffic and congestion near the school but also advanced more general objectives for community development and urban sustainability.

16.2.3.3 Tools and Methods Applied

The project's execution in Xianning made use of a number of cutting-edge instruments and techniques intended to tackle the particular difficulties associated with traffic safety and urban planning near schools. The use of a human-centered design, in which the needs of the students were the primary consideration in all decisions, was a fundamental component of the strategy. In order to make the walkways both safe and practical for kids, they were especially designed for students' commutes. Child-friendly elements, such as suitable lighting, unambiguous signage, and barriers that shield young pedestrians from traffic, were emphasized in the design of these pathways. Children now have a safer and healthier way to get around thanks to the redesign of the streets and sidewalks, which created a friendly atmosphere that promoted walking instead of driving.

The project concentrated on developing areas that would fulfill both practical and social functions in addition to the infrastructure upgrades. The introduction of a co-created garden turned underutilized or unused areas into lively green spaces that the school community could enjoy. Students, educators, and locals all contributed to the garden's design, which made sure it satisfied the community's various needs. In addition to being a place for students to unwind and have fun, this common area developed into a first-rate community resource where locals could interact with one another. The project's overarching objectives of establishing a child-friendly urban setting were all aided by the garden's promotion of social interactions, enhancement of mental health, and sense of group ownership over the area.

Several Sustainable Development Goals (SDGs) served as inspiration for the project, which was also firmly grounded in sustainable development principles. The project's planning and implementation were in line with the SDGs, which emphasize creating more sustainable, safe, and inclusive cities and communities. The initiative's focus on establishing green areas and pedestrian-friendly walkways directly aided in achieving SDG 11, which calls for the creation of sustainable cities. Because materials were recycled and attempts were made to lower the carbon footprint of construction by using eco-friendly materials like recycled tires, metal, and bottles, the environmental sustainability aspect was especially significant. Furthermore, the garden itself served as a vehicle for advancing environmental education by providing students with practical experience in sustainability.

In addition to successfully addressing the immediate traffic safety issues surrounding schools, the initiative advanced a long-term vision for more sustainable and community-oriented urban spaces by fusing sustainable urban planning principles with a human-centered approach.

16.3 Obstacles and Solutions

Numerous obstacles had to be overcome during the implementation of the initiative, some of which called for innovative thinking and creative solutions. The requirement for more money to support the redesign of infrastructure was one major barrier. The project was ambitious and included many innovations, including the construction of green spaces, new pedestrian pathways, and a restructure of traffic patterns around schools. Despite being necessary, these adjustments came with higher costs that weren't fully factored in at the beginning. As a result, the project received strong support from the city's leadership and pertinent decision-makers, who saw its potential long-term advantages. Departmental budgets were modified as a result of this support, with money taken from other initiatives to pay for the higher costs.

The unconventional and distinctive structures that the students created presented another difficulty because they had to adhere to strict engineering specifications despite their inventiveness. When it came to turning their imaginative concepts into tangible items, the student-designed components—such as street furniture and decorations for public areas—presented a technical challenge. In order to get around this, techniques like 3D printing and handcrafting with fiberglass were employed, guaranteeing that the finished structures adhered to recognized safety standards and engineering practices. These solutions met the pragmatic needs of urban infrastructure while also maintaining the integrity of the students' creative contributions.

The substantial amount of recycled materials used in the project's construction presented another difficulty. Despite the initiative's dedication to sustainability, the use of materials like metal, tires, and used bottles created issues for the engineering acceptance and auditing procedures. Because it was perceived as a departure from standard building practices, the use of recycled materials was frequently criticized. The initiative developed new city-level design guidelines in close collaboration with the urban planning bureau to address these issues. By establishing a standard for future initiatives and guaranteeing that the use of recycled materials complied with local building codes, these guidelines made it clear how these materials could be used in urban projects in a safe and efficient manner.

Notwithstanding these challenges, the project's creative solutions showed resiliency. In the end, the initiative overcame the difficulties encountered during its implementation thanks to the cooperation of city officials, departments, private developers, and community members. The project was able to realize its goal of creating safer, more sustainable school environments in Xianning thanks to modifications made to the budget, engineering procedures, and material usage.

16.3.1 Achieved Outcome

Traffic safety and urban planning around Xianning's schools have significantly improved as a result of the initiative's implementation. The distinct division between students walking to school and those dropping off private vehicles was one of the most noteworthy results, as it successfully reduced traffic on the school grounds. About 30% of students walked to school prior to the initiative, and private vehicle drop-offs added to the gridlock. However, the percentage of students who walk to school soared to an astounding 95% after pedestrian-friendly pathways and other infrastructure improvements were implemented. Alongside this change, there was a sharp decline in the number of private vehicles used to drop off students.

Traffic flow improved and pedestrian conditions, particularly for children, became much safer because of the sharp decline in the number of cars from 120 to just 8 per hour. The sharp decline in traffic accidents near the school was another advantageous result. Approximately fifty traffic incidents occurred annually within a one-kilometer radius of the school prior to the initiative. But after the initiative's modifications were made, this number fell to just one accident annually, and that one had nothing to do with children. This significant drop in accidents is evidence of the effectiveness of the redesigned infrastructure and the emphasis on making the school and the surrounding areas safer for children.

The 800 students who attended the 7th Elementary School, their families, and nearby residents were all directly impacted by the initiative, as they benefited from increased safety and the development of more accessible and community-focused areas. The program promoted a more involved and cohesive community in addition to addressing the pressing issues of safety and traffic congestion. Due to the success of the initial implementation, the school environment renovation became a model for similar projects throughout the city, and 20 more schools in Xianning have replicated the initiative. More than 30,000 elementary school students and preschoolers now benefit from safer school commutes and improved access to public areas, with these schools sharing in the advantages.

Beyond the actual infrastructure improvements, the initiative has had a significant cultural impact by promoting more environmentally friendly and child-friendly urban planning. The project has established a good example for other Xianning cities and schools by attending to the needs of students, families, and the community at large, demonstrating that such programs can greatly improve everyone's safety and quality of life.

16.4 Lessons Learned

The Xianning initiative's success offers important insights into urban development and the significance of incorporating child-friendly ideas into city planning. Cities all over the world have struggled to revitalize urban areas that have suffered from a decline in social interactions and economic activity since the COVID-19 pandemic. The significance of improving urban areas' resilience—not only economically, but also in terms of social, environmental, and infrastructure sustainability—is timely brought to light by this project. Through its emphasis on enhancing traffic safety and developing first-rate public areas for children, the project showed how important these kinds of initiatives can be in promoting urban vitality and a feeling of community.

Understanding and taking into account the needs of children in urban development is one of the most important lessons this initiative has taught us. The ability of the project to actively involve children in the design and decision-making process was a major factor in its success. The initiative made sure that the built environment satisfied the needs of the students by approaching the project from a "one-meter height" perspective and paying attention to their ideas. In addition to addressing immediate safety concerns, this method assisted kids in feeling a sense of connection and ownership over the area. Students were able to express their creativity, gain confidence, and develop critical social skills by working together to co-create elements like public furniture and gardens. Thus, the program established a strong basis for their future growth, highlighting the long-term advantages of developing these abilities at a young age.

The project also demonstrated the importance of teamwork in achieving urban sustainability. This project showed that genuine sustainability and growth necessitates the cooperation of a wider range of actors, whereas traditional urban development frequently involves a small number of stakeholders, such as local governments and construction companies. Parents, members of the community, nearby companies, and even private developers were successfully involved in the project and all contributed significantly to its success. This inclusive approach made sure that the results were in line with the city's larger social and environmental objectives in addition to being pertinent to the school's immediate needs. The focus on "co-funding, co-building, and co-creating" in the initiative served as a template for other cities to adopt. It illustrated how an emphasis on people can propel urban change, with neighborhoods cooperating to improve their environment and standard of living.

The importance of low-investment, quickly implementable solutions with high promotional value is another crucial lesson to be learned from this initiative. The project had significant effects despite being finished quickly and requiring only a small amount of funding. This demonstrates that cities don't need big budgets or long timelines to accomplish transformative outcomes. The project's success in making the school a child-friendly place while addressing traffic safety further demonstrated that small changes can have a significant and long-lasting effect on the neighborhood. With an emphasis on making effective use of the resources at hand to create areas that are not only safe but also lively, inclusive, and engaging for all citizens, it provides a replicable model for other cities dealing with comparable issues.

Lastly, the project emphasizes how public space can promote a feeling of pride and community ownership. The project promoted a sense of collective ownership by incorporating different stakeholders in the development of public spaces and motivating them to assume accountability for preserving and enhancing those areas. Involving the public, particularly children, in the development of their surroundings fosters a sense of pride and belonging that supports the area's long-term viability. With this project, Xianning has developed a model that not only tackles current urban problems but also creates a more resilient, inclusive, and child-centered urban environment, providing other cities aiming for comparable objectives with important insights.

Related SDGs: Goals 3, 4, 5, 11, 13, 15, and 16.

Acknowledgements We explicitly thank the project owners based in Xianning, China, for their participation and the provision of materials for the 6th Guangzhou International Award for Urban Innovation. We also congratulate them on their achievements in the 2023 cycle of awards.

The above images are provided to us by the project owners for their submission.

The manufacturer's authorised representative in the EU is Springer Nature Customer Service Centre GmbH, Europaplatz 3, 69115 Heidelberg, Germany. If you have any concerns regarding our products, please contact ProductSafety@springernature.com

Printed and bound by CPI Group (UK) Ltd, Croydon, CR0 4YY
01/10/2026
02225929-0003